Bristol Radical Pamp

Turbulen

Labour and Gender Relations in Bristol's Aircraft Industry during the First World War

Mike Richardson

ISBN 978-1-911522-42-3
Copyright © with the author Mike Richardson 2017

Bristol Radical History Group. 2017.
www.brh.org.uk
brh@brh.org.uk

Contents

Acknowledgements iv
Picture Credits v
Introduction 1
The British & Colonial Aeroplane Company 1910-14 1
British & Colonial Aeroplane Company in the early months of the war 5
1915 Sources of Conflict 9
Craft workers' reaction to the Munitions Act 15
Bristol workers' response to dilution and conscription 20
Conscription, the labour supply and the increase in demand for women workers 24
Labour Unrest: Class and Gender Divisions 29
Labour Relations: the State, employers, trade unions and workers' representatives 34
The Armistice 48
Bristol Aircraft Companies in 1920 51
Summary and Conclusion 53
Appendix 57
Bibliography 64
Name Index 69
General Index 70

Acknowledgements

Thanks, firstly, to Sheila Rowbotham, comrade and partner, for reading and editing several drafts of this work. Thanks also to Di Parkin for proof reading, Richard Grove for the cover design and the preparation of the final text for print production, and to the Bristol Radical History Group for their continued support to write and publish history from below. I am indebted to The Bristol Central Reference Library, the Bristol Archives, Bristol Museums Galleries Archives and the University of the West of England for the use of their facilities.

Images/Text are reproduced with the kind permission of The British Newspaper Archive (FindmyPast Newspaper Archive Limited), (www.britishnewspaperarchive.co.uk), and Trinity Mirror plc. For quotes and references to nineteenth century British newspapers; Text © THE BRITISH LIBRARY BOARD. ALL RIGHTS RESERVED. In respect to the *Western Daily Press and Bristol Times and Mirror*, text are reproduced with the kind permission both The British Newspaper Archive and the Bristol Central Reference Library; photographs, courtesy of Bristol Archives, Bristol Museums Galleries Archives and my grandson Joshua Clay.

Picture Credits

Page vi: Bristol Museum: Box-Kite Replica, photograph by Joshua Clay

Page 2: Bristol Biplane First Flight from Clifton Downs, Bristol Archives, 43207/10/2/2

Page 4: Bristol Scout B, Bristol Archives, 42421/Ph/6

Page 17: Women working on the construction of Bristol F2B Fighter: Wood Details, Bristol Museums Galleries Archives, J4747.5

Page 22: Men working on the construction of Bristol F2B Fighter: Sheet Metal Shop, Bristol Museums Galleries Archives, J4747.6

Page 28: Women & Men working on the construction of Bristol F2B Fighter: Doping Wings, Bristol Museums Galleries Archives, J4747.4

Page 35: Strachan & Henshaw, Women Munitions Workers 1918, Bristol Archives, 40824/3

Page 39: Women & Men working on the construction of Bristol F2B Fighter: Page Assembling Planes, Bristol Museums Galleries Archives, J4747.2

Page 40: Women & Men working on the construction of Bristol F2B Fighter: Cabinet Shop, Bristol Museums Galleries Archives, J4747.1

Page 41: Women Workers at Parnall's Coliseum Aircraft Works, 1918, Photograph by E. C. Stevens, Bristol Archives, 43207/9/43/30

Page 50: British & Colonial Aeroplane Company, Brislington Workers, 1918, Bristol Archives, 39735/IM/Ph/18/16

Page 52: Women & Men working on the construction of Bristol F2B Fighter: Final Erecting, Bristol Museums Galleries Archives, J4747.3

Appendix Page 57: Entwistle: Army Reserve Scheduled Occupation Certificate, Bristol Archives Bristol Archives, 37267/1 to 10

Box-Kite replica, Bristol Museum.

Introduction

In the early twentieth century Bristol, which already had a tradition of manufacturing, was quick to make its mark as a pioneer centre for the aircraft industry. The first business in Bristol and its environs to enter this new industry was the British & Colonial Aeroplane Company, established in 1910 by Sir George White, the chairman of Bristol Tramways. A replica of the company's first major undertaking of the light-built, skeletal structured, airy Box-Kite aircraft still hangs in the front hall of the Bristol City Museum and Art Gallery. The outbreak of the First World War brought two more Bristol companies into manufacturing planes and parts for military aviation. Parnall and Sons applied its expertise as shopfitters, iron and brass founders, and the manufacturer of scales and weighing machines, to aircraft production. Then in 1915, the engineering and motor manufacturers, Brazil Straker & Company Ltd., entered the industry using its know-how and experience to produce aero engines.[1]

By 1910 air displays were attracting tens of thousands of people to witness the wonder and excitement of flying machines.[2] Demonstrations of flights in Bristol took place on Clifton Down with crowds of several hundred attending.[3] Although between 1910 and 1914 only 863 people learnt to fly in Britain,[4] the early planes and their pioneering pilots were to pass into popular culture through specialist magazines, eventually figuring in comics and becoming best selling children's books.[5] Less visibly, the early aircraft industry itself was a site for equally dramatic developments before and during the war. It saw private investment, government support via State guarantees, State regulation, new production methods, changes in workforce composition and labour relations. The tensions which resulted mean aircraft production can be regarded as a microcosm of the wider impact of war on the armaments industry.

The British & Colonial Aeroplane Company 1910-14

In 1910, the Secretary of State for War, Richard Haldane (from 1911 Viscount Haldane) granted the British & Colonial Aeroplane Company (hereafter referred to as B&CAC) flying rights, for carrying out flight-testing, on just

---

1 Grace's Guide to British Industrial History, gracesguide.co.uk accessed 1 January 2017.
2 L. C. S. Budd, 'Selling the Early Air Age: Aviation Advertisements and the Promotion of Civil Flying in Britain, 1911-1914', *Journal of Transport History*, 32. 2 (Dec 2011), p. 125.
3 *Western Daily Press*, 12 November 1910, p. 5.
4 Royal Air Force Museum, rafmuseum.org.uk accessed 27 April 2017.
5 In America, M. A. Donahue & Co. published a series of books for both boys and girls on aviators and aeroplanes before the First World War, see M. Cadogan, *Women with Wings: Female Flyers in Fact and Fiction* (Macmillan, 1992), p. 40.

Bristol Biplane first flight from Clifton Downs, 1910.

under three acres of land at Lark Hill, Salisbury Plain.[6] In 1911 George White converted his Filton tramway depot, built in 1907 on the two acres of land he had acquired next to the site of his old home, into an aircraft factory and airfield.[7] His B&CAC set about recruiting men with engineering, metal and woodworking skills, that could be easily adapted to the construction of aircraft, from around the country as well as from Bristol's existing workforce. In March 1911, a couple of months after the appointment of George White's son, G. Stanley White as the company's managing director, the War Office placed an order for four 'Bristol' Military Biplanes (Box-Kites).[8] In the same year the Admiralty called upon the company to develop, in secret, an inflatable hydroaeroplane that when deflated could be stored aboard a submarine.[9] However, with the cost at £3,200 for each machine, and some technical problems, the company decided not to proceed, and in 1914 it abandoned the project.[10]

6 The British and Colonial Aeroplane Company Minute Book, No. 1 (Royal Aeronautical Society Archives, RAES/BCAC/DIRE/1), 24 May 1910.
7 C. H. Barnes, 'Bristol and the Aircraft Industry', *Bristol Industrial Archaeological Society Journal* 34, 1972, accessed online 2 January 2017, www.b-i-a-s.org.uk.
8 The British and Colonial Aeroplane Company Minute Book, No. 1, 18 January & 15 March 1911; T. C. Treadwell, *British & Allied Aircraft Manufacturers of the First World War* (Amberley, 2011), p. 78.
9 J. M. Bruce, 'The Bristol Scout: Historic Military Aircraft No. 18, Part 1', in *Flight*, 26 September 1958, pp. 525-6.
10 The British and Colonial Aeroplane Company Minute Book, No. 1, 26 June 1914.

The close association between employers and the State was not peculiar to Bristol, since aircraft manufacturers depended on government contracts to sustain and expand their enterprises. Moreover, in the edgy years just before World War One, aircraft production was lagging behind that of Germany, so the State turned to private sector companies, like B&CAC, to make up the shortfall.[11] Hence a symbiosis between government and business emerges. In 1913, the Director-General of Military Aeronautics took over responsibility for the procurement of aircraft and associated parts from the director of Army Contracts.[12] The State already had direct control over the research and design of aircraft at its Royal Aircraft Factory based in Farnborough, a few of which came to be mass-produced during the war by firms, including B&CAC, in the civil sector. As the drift towards war began to feel inevitable, new and existing firms started to respond to the increased demand for military aircraft.[13]

In July 1914, B&CAC received orders from the Admiralty for four B.E.10 biplanes and eighteen B.E.2c reconnaissance aircraft, a biplane design of the Royal Aircraft Factory. More orders followed in early August, after Britain had declared war on Germany, for fourteen Bristol Tractor biplanes and a further twenty-eight B.E.2c.[14] The company responded immediately by recalling its employees from their summer holidays and putting them to work on the contract. But errors found in the technical drawings, supplied by the Royal Aircraft Factory for the B.E.2c aircraft, delayed production and the Admiralty turned its attention to the company's single-seater biplane, known as the Bristol Scout. Two were ready for a final overhaul and some minor modifications. The Admiralty requisitioned these and, after the necessary alterations and servicing, they were delivered to the Royal Aircraft Factory, at Farnborough, in the third week of August.[15]

The rise of the aircraft industry coincided with rapid technological advances which disrupted established custom and practices at work. At the same time, between 1911-14, a mood of confidence flowed from an upsurge of rank-and-file militancy among miners, railway men, dockers and building workers who not only demanded improvements in pay but also sought greater

---

11 J. D. Murphy, 'Aircraft Production' in S. C. Tucker (ed.) *World War 1 Encyclopedia* (California: ABC-CLIO, 2005), p. 57.
12 *The Official History of the Ministry of Munitions*, Vol. 1, Pt. 1, Ch. 11 (HMSO, 1922), p. 50.
13 *The Official History of the Ministry of Munitions*, Vol. V111, Pt. 1, Ch. 1 (HMSO, 1922), p. 30.
14 The British and Colonial Aeroplane Company Minute Book, No. 1, 15 September 1914; *The Official History of the Ministry of Munitions*, Vol. X11, Part 1, Ch. 1, (HMSO, 1922) p. 23.
15 The British and Colonial Aeroplane Company Minute Book, No. 1, 7 August and 15 September 1914; Bruce, 'The Bristol Scout:' 26 September 1958, p. 527.

Bristol Scout B.

control over their changing conditions of labour.[16] Although the rebellions were mainly in male industries, women factory workers were also caught up in eruptions of resistance. They were organised in the main by the National Federation of Women Workers, and the Workers' Union which included both male and female unskilled and semi-skilled workers.[17]

Neither new technology nor labour militancy, however, dislodged an authoritarian approach to industrial relations which persisted from the nineteenth century. Employers tended to assume that their power as owners of the means of production should be absolute and took for granted their right to dictate the terms of employment, the hiring and firing of labour, and how work was carried out. Such assumptions of managerial prerogative accentuated unrest among employees. Friction arose over the right to unionise, the subdivision of manufacture, opposition to dilution and the use of semi-skilled and unskilled workers, particularly women, to do the work of skilled craftsmen.

The outbreak of war transformed the context but did not resolve these conflicts. Indeed they intensified. The rush of young men to sign up for military service, the hasty introduction of the Defence of the Realm Act (4 August 1914)

---

16 For an account of unrest in this period see H. A. Clegg, *A History of British Trade Unions since 1889: Volume 11 1911-1933* (Oxford University Press, 1985), pp.24-74. For Bristol's experience of industrial strife see M. Richardson, 'Bristol and the Labour Unrest of 1910-14' in D. Backwith, R. Ball, S. E. Hunt and M. Richardson (eds.) *Strikers, Hobblers, Conchies & Reds: A Radical History of Bristol 1880-1939* (London: Breviary Stuff Publications, 2014) chapter 6, pp. 197-238.

17 See S. Rowbotham, *Hidden from History* (Pluto, third edition, 1977) pp. 108-11. For a strike of textile workers supported by the Workers' Union see M. Richardson, *Bliss Tweed Mill Strike 1913-14: Causes, Conduct and Consequences* (Bristol Radical History Group, Pamphlet 26, 2013).

and the patriotic fervour displayed by many union leaders posed a threat to the hard-earned gains workers had won in the pre-war era. What happened in Bristol is thus a crucial part of the larger national picture.[18]

British & Colonial Aeroplane Company in the early months of the war

In September 1914, after Germany had invaded France, two modified Bristol Scouts were sent to the combat zone to assist the French in pushing back the Germans at the battle of the Marne. The Scout's performance during this operation won the plane the nickname of the Bristol Bullet. Further orders followed. In October, the Admiralty purchased two 'Bristol' Tractor biplanes and six 'Bristol' School biplanes.[19] Not all was going smoothly, however. The problems with the technical drawings in regard to the B.E.2c aircraft were still unresolved. In November, the B&CAC wrote to the War Office remonstrating that the drawings supplied by the Royal Aircraft Factory were inaccurate and requested that the errors should be rectified immediately.[20] Until this was done it could not proceed with production of these aircraft. The Royal Aircraft Factory sent a draftsman to the company to do the necessary corrections enabling the first B.E.2c to be delivered on 16 December.[21] In the meantime, the War Office ordered a further fifty B.E.2c's and twelve new Bristol Scouts. A single rotary eighty-horse power piston engine propelled the well-constructed Bristol Scout making it faster, sturdier and more stable than earlier models of the biplane. The fuselage was made of a wire-braced wooden structure (seasoned ash) with a fabric covering at the rear and sheet aluminium at the forward section. The wings were also fabric-covered.[22] They were built at the Brislington works of the Tramways Company of which Sir George White was chairman and a major shareholder.[23]

---

18 While Bristol has received relatively little attention studies of the national picture have proliferated. For examples see B. Drake, *Women in the Engineering Trades* (Fabian Society & George Allen & Unwin, 1917); B. Drake, *Women in Trade Unions* (Virago, 1984, first published in 1920 by the Labour Research Department); G. Braybon, *Women Workers in the First World War: The British Experience* (Routledge, 2013, first published 1981); A. Woollacott, *On Her Their Lives Depend: Munitions Workers in the Great War* (University of California Press, 1994); J. T. Murphy, *Preparing for Power* (Pluto Press, 1972, first published by Jonathan Cape, 1934); J. Hinton, *The First Shop Stewards' Movement* (London: George Allen & Unwin, 1973).
19 The British and Colonial Aeroplane Company Minute Book, No. 1, 13 October 1914.
20 The British and Colonial Aeroplane Company Minute Book, No. 1, 30 November 1914.
21 The British and Colonial Aeroplane Company Minute Book, No. 1, 28 December 1914.
22 *Flight*, 25 April 1914, p. 433.
23 The British and Colonial Aeroplane Company Minute Book, No. 1, 30 November 1914; Bruce, 'The Bristol Scout', p.528.

Having established a reputation for creativity, design, engineering and craft skills, B&CAC, in order to sustain the momentum it had achieved, turned its attention to the recruitment and retention of skilled workers. In early January, to incentivise existing staff, and attract new staff, the company increased wage rates of the total workforce by five per cent. Moreover, in a notice to its employees it announced that the directors were considering awarding a productivity bonus 'as a further encouragement in respect of "the machines to be turned out."'[24]

Following the hike in wages, advertisements began to appear in the local press, the first at the end of January 1915, for a wide range of skilled workers including carpenters, motorcar and coach body builders, tinsmiths and sheet metal workers, fitters, turners, millers, trimmers (to apply fabric to aircraft wings) and draughtsmen.[25] Notably the B&CAC only invited men to apply. Like the majority of other employers it indicated that it wanted skilled men rather than women. As one of the founders of the Labour Department of the Ministry of Munitions, Humbert Wolfe, remarked 'the prejudice against [hiring] female labour could not be killed by three months of war.'[26] Steps to introduce the dilution of labour (the breakup of craft work into its simplest elements and the relaxation of demarcation restrictions to allow semi-skilled or female labour on munitions work)[27] were still at an early stage. At this time there was little pressure on the Amalgamated Society of Engineers (ASE), the National Amalgamated Sheet Metal Workers and the Amalgamated Society of Carpenters and Woodworkers, unions representing the majority of the craftsmen working in the aircraft industry, to allow the introduction of female labour into their trades. This was about to change as the army increasingly demanded measures from the government to secure the release of men from industry to replace those fallen in battle.

The orders for the new Bristol Scout, designated Bristol Scout C, were delivered in March 1915, following which the Royal Flying Corps called for a further seventy-five and an additional one hundred and fifty B.E.2c biplanes.[28] The company also received a requisition from the Scottish engineering firm, G. & J. Weir, for 1,000 wood propellers for aircraft engines destined to go to the Royal Aircraft Factory.[29] Again in March 1915 the B&CAC raised wages, this time by 2s 6d per week (12½ pence) making the increase in wages since the beginning of the year equivalent to just over 10 per cent (inflation was

24 The British and Colonial Aeroplane Company Minute Book, No. 1, 14 January 1915.
25 For example see *Western Daily Press*, 29 January 1915, p. 2 and 23 April 1915, p. 2.
26 H. Wolfe, *Labour Supply and Regulation* (Oxford: Clarendon Press, 1923), p. 79.
27 See H. Braverman, *Labor and Monopoly Capital* (Monthly Review Press, 1974).
28 Bruce, 'The Bristol Scout', p. 528.
29 The British and Colonial Aeroplane Company Minute Book, No. 1, 2 March 1915.

running at 12½ per cent in 1915).[30] However, this pay award did not quell the ambition of craft unions in Bristol to strengthen their bargaining position in the swiftly developing aircraft industry. Attracting men from across the country, as well as in the local area, the B&CAC labour force grew from 400 in 1911 to 1,000 strong by the middle of 1915,[31] bringing in its wake a call for union recognition.

The war created an ethos of patriotism which spread to the leadership of the unions. They were prepared to work with the employers and the State in ways that transformed how industrial relations were conducted. Trade union leaders' co-operation with and support for the war effort had resulted, in March 1915, in a voluntary accord between the government, employers' federations and trade unions, known as the Treasury Agreement. If its objectives had been realised, this would have brought about the suspension of strikes, lock-outs and output restrictions for the duration of the war along with the beginnings of dilution.[32] An initial consequence of the Treasury Agreement was a tactical shift and a change of attitude on the part of some employers. Many firms previously hostile to conceding union recognition felt that it was in their best interest to modify their opposition to trade unions.

In Bristol, a clear example of this new forbearance came about in April 1915 when, despite Sir George White's prevailing anti-union sentiments, the B&CAC conceded union recognition to its skilled male workforce. On securing accreditation, the Sheet Metal Workers' Union took advantage of its newly acquired bargaining strength by immediately mandating its members not to accept work at B&CAC at less than a wage of 42/- per week.[33] Moreover, two sheet metal workers who had started work at White's Brislington depot in May for 9 pence an hour were instructed by their union not to work for less than 9½ pence an hour.[34] The union's push for higher rates of pay proved to be timely.

On Saturday 12 June 1915, the newly appointed Minister of Munitions, David Lloyd George MP, arrived in Bristol on his whistle stop tour of large industrial centres in Britain. His visit included an inspection of the B&CAC Works at Filton. After observing the processes involved in manufacturing aircraft he addressed the workforce, thanking them for their contribution to the

30 The British and Colonial Aeroplane Company Minute Book, No. 1, 22 March 1915.
31 *Western Daily Press*, 21 June 1915, p. 9.
32 *The Official History of the Ministry of Munitions*, Vol. 1, Pt. 2, Ch. IV (HMSO, 1922), pp. 85-88.
33 Minutes of the General Meeting of the Bristol Operative Tin and Iron Plate, Sheet Metal Workers and Braziers, 26 April 1915, Mike Richardson papers.
34 Minutes of the General Meeting of the Bristol Operative Tin and Iron Plate, Sheet Metal Workers and Braziers, 8 July 1915.

war effort but emphasising that 'we want more aeroplanes…The more machines you can turn out the better it will be for the men at the front.'[35]

Taking note of Lloyd George's plea to Bristol's aircraft workers, three days later the B&CAC announced the introduction of a 'Special War Bonus' scheme, the details of which were posted in the workshops:

> the Directors have now decided that a Bonus of Twenty Pounds per Machine [aircraft] shall be set aside and divided amongst the workmen (*sic*) who are working for the Company at the end of the War. (This will include the men engaged on aeroplane work at the old Brislington Tramway Department and the "Scouts" being turned out there will come into the general reckoning).[36]

Backdated to the beginning of the year this bonus had already accrued a sum of around £8,000. However, the directors included one important caveat: 'each man must clearly understand that he must see the job through to the end of the War and no man leaving for any cause whatever would participate' in the bonus scheme. This was expected to amount 'approximately to 10 per cent on the wages paid during the period of the War'. A bonus of £2 per machine was granted for division among the foremen, drawing office and clerical staff on the same qualifying terms as the production workers.[37] In its push to meet existing contracts and win more lucrative aircraft work the company had devised an incentive reward scheme shaped specifically to retain its current workforce, recruit new staff and up production.

The company had already entered into talks with the deputy director of Military Aeronautics on building an extension to its factory in return for more orders.[38] And in July 1915, it received a positive response from the War Office to its request that in return for expanding its capacity it would receive contracts to produce three hundred and fifty B.E.2 D aeroplanes and fifty 'Bristol' Scouts and that it should 'be kept fully occupied to the end of 1916'.[39] The Bristol Tramways Company, in which George White was a major shareholder, won the contract to carry out much of the extension work, including the electrical power supply.[40]

---

35 *Western Daily Press*, 14 June 1915, p. 6.
36 The British and Colonial Aeroplane Company Minute Book, No. 1, 16 June 1915.
37 The British and Colonial Aeroplane Company Minute Book, No. 1, 16 June 1915.
38 The British and Colonial Aeroplane Company Minute Book, No. 1, 16 June 1915.
39 The British and Colonial Aeroplane Company Minute Book, No. 1, 7 July 1915; H. Driver, *The Birth of Military Aviation: Britain, 1903-1914* (Boydell & Brewer, 1997), pp. 120-1.
40 The British and Colonial Aeroplane Company Minute Book, No. 1, 16 June and 7 July 1915.

1915 Sources of Conflict

Despite the pressure for industrial peace, by the spring of 1915 tensions were already appearing around the need for skilled workers and men for the military. These gave rise to exploratory methods aimed at finding the most efficient and effective way of redistributing labour. As a consequence, in May 1915 the North-East Coast Armaments Committee launched a trial of the King's Squad Scheme (a Flying Column of Armament Workers). This voluntary scheme was organised by a local committee, comprising government officials, employer representatives and workmen. It proved to be attractive and simple.

Men who volunteered for the King's Squad gave their consent to go to any workshop or locality engaged in munitions, where their services were required. In return, every workman thus transferred 'would earn the same (or more) wages and be under no military restrictions whatever' and be entitled to subsistence or travelling allowance.[41] This was an appeal to workers to leave commercial work for armament work and 'gave protection during 1915 against gibes of slackness and in 1916 against military service.'[42] Though workers thus gained freedom of movement, once having taken this option *they would have to give up their right to choose where they wished to work*. The government believed this would constitute an advantage to munitions employers whose employees signed up to the scheme. By joining men could not leave of their own accord so a stable workforce could be ensured.[43]

The King's Squad Scheme produced promising results which convinced the government, in June 1915, to roll out a comparable plan nationwide, called the War Munitions Volunteers' Scheme. To all intents and purposes, this 'industrial army' scheme was a national expansion of the King's Squad Scheme it replaced. It operated along similar lines and continued alongside conducting recruitment to the armed services on a voluntary basis. Most significantly, however, it made accessible certain classes of skilled labour that previously had been difficult to obtain for munitions production. This applied particularly to those skilled men who had enlisted in the armed forces and were now no longer restricted by their military superiors from returning to civilian life as munitions workers.[44] The War Munitions Volunteers' Scheme was launched with great fanfare and publicity. On 14 June, the Director General of Military Aeronautics, David Henderson, contacted the B&CAC advising that as a result of the new scheme its 'employees should be enrolled under Military control.'[45] And on 2 July, the

---

41 *The Official History of the Ministry of Munitions*, Vol. 1., Pt. 3, Ch. 111, p. 46.
42 Wolfe, *Labour Supply and Regulation*, p. 108.
43 *The Official History of the Ministry of Munitions*, Vol. 1., Pt. 3 Appendix XIV, p. 132.
44 D. Lloyd George, *War Memoirs of David Lloyd George*, Vol. 1 (Odhams Press, 1938), pp. 173 & 182-3.
45 The British and Colonial Aeroplane Company Minute Book, No. 1, 16 June 1915.

Director of Air Organisation, Colonel Brancker, visited both the Filton and Brislington Works and distributed War Badges and War Medal Certificates to B&CAC skilled workers who had signed up to the Scheme.[46] Across the country over one hundred thousand enrolled but few were transferred because most of the volunteers were already employed on work regarded by the Ministry of War as indispensable to the war effort.[47]

Employers did not regard all aspects of government intervention with favour. On 7 July, at a meeting of the company directors, the managing director of the B&CAC, G. Stanley White, informed his fellow directors that the Minister of Munitions, Lloyd George, had notified him by letter that the he intended 'to declare the Company's establishment a controlled establishment', as the company was already producing munitions of war.[48] White and his directors did not take too kindly to outside interference into the running of their company. They were great advocates of full and open competition and were concerned that the government would reorganise the aircraft industry as a State monopoly. Nonetheless, despite the company's objections, on 18 August the official proclamation that the company had been declared a 'controlled establishment' arrived.[49] The receipt of government orders, for fifty more Bristol Scouts, twenty-four Bristol School biplanes, two hundred B.E.2.Ds and five hundred aviators' safety belts, compensated for this unwelcome news.[50] The unprecedented demand for aircraft upped the pressure to set in train dilution and the employment of women and unskilled men on aircraft work.

This process was accentuated because significant numbers of skilled workers had enlisted in the army, depleting the numbers available to work in the aircraft industry and elsewhere in the munitions factories — Albert Edward Clarke is one example of a skilled woodworker who joined the armed services during the war. He lived within a few miles of the Filton aircraft works and was probably employed there.[51] Paradoxically, the new conditions put skilled

---

46 The British and Colonial Aeroplane Company Minute Book, No. 1, 7 July 1915.
47 R .J. Q. Adams, 'Delivering the Goods: Reappraising the Ministry of Munitions 1915-1916', *Albion: A Quarterly Journal Concerned with British Studies*, Vol. 7, No. 3, Autumn, 1975, p. 240.
48 The British and Colonial Aeroplane Company Minute Book, No. 1, 7 July 1915.
49 The British and Colonial Aeroplane Company Minute Book, No. 1, 19 August 1915.
50 The British and Colonial Aeroplane Company Minute Book, No. 1, 19 August, 11 November 1915.
51 Clarke moved from Poole, in Dorset, during 1915, to 108 Bishop Road Horfield. (Census of England and Wales 1911, courtesy of the National Archives/Findmypast Limited 2015 and Bristol Street Directories, 1915-20, (Matthews, Wrights and Kelly's), Bristol Central Reference Library). In a letter to the editor of *The West Bristol Labour Weekly*, 29 October 1926, James Corbett referred to Clarke 'as a well-known ex-serviceman' standing in the local election for Labour in the Horfield Ward in 1926. However, the 'badge of honour' of being an ex-serviceman was not enough to get him elected.

aircraft workers in a stronger bargaining position, even while placing them under pressure to relax their lines of demarcation. Doggedly, they persisted in resisting the introduction of semi-skilled and unskilled workers, particularly women, on work that they had customarily done. In some instances they were successful. Sheet metal workers were able to maintain their skilled status and union conditions largely through the action of rank-and-file union members in their workshops. A 'proposal by Farnborough authorities [north east Hampshire] to put women on cutting out was quietly shelved when the shop organised a protest meeting.'[52]

By July 1915 it had become apparent that the March Treasury Agreement had failed to deliver, as union members in many areas of the country were less willing than their leaders to abide by its stipulations. This failure, along with the scandal that had broken out over the shell-shortage crisis, forced the government, with the assent of employers and union leaders, to pass the Munitions of War Bill 1915, which became law on 2 July. It gave power to the Ministry of Munitions to proclaim as a controlled establishment any factory or workshop whose business resided 'wholly or mainly in engineering, shipbuilding, or the production of arms, ammunition or explosives, or the substances required for the production thereof.'[53] Thus attempts at voluntarism were abandoned in favour of compulsion.

In recalling this decision to resort to compulsion, the Ministry of Munitions official, Humbert Wolfe, maintained that

> Neither the Trade Unions nor the working men were opposed, when all else had failed, to compulsion for strictly military purposes. But, though they might be willing to be conscripted to die for an idea, they were not to be conscripted to live (as it might have been put) for private profits.[54]

The aircraft industry was doing very well out of the war which brought its business leaders round to working closer with the government,[55] despite the initial reticence of some, such as Sir George White. In reality, unitary corporatism had emerged. In the words of Robert Currie this was a 'system devised for war work [that] denied the possibility of conflict of interests and vested overall industrial control, within each establishment, in the hands of employers.'[56]

52 T. Brake, *Men of Good Character: A History of the Sheet Metal Workers, Coppersmiths, Heating and Domestic Engineers* (Lawrence and Wishart, 1985), p. 149.
53 *The Official History of the Ministry of Munitions*, Vol. V, Pt. 1, Ch. 11, p. 34.
54 Wolfe, *Labour Supply and Regulation*, p. 42.
55 B. W. Alford, *Britain in the World Economy Since 1880* (Longman, 1996), p. 111.
56 R. Currie, *Industrial Politics* (Oxford: Clarendon Press, 1979), p. 92.

Initially, public opposition to the 1915 Act was surprisingly constrained given that it gave the newly created Ministry of Munitions greater powers than those contained in the March Agreement.[57] However, John Maclean, revolutionary socialist and founder member of the Clyde Workers' Committee, formed in 1915, perceived the drastic implications this law entailed, branding it 'the Industrial Slavery Act'.[58] Indeed it handed the Ministry of Munitions the legal authority over civil munitions works, known as 'controlled establishments', to regulate wages as well as rendering strikes and lock-outs illegal. The Ministry, under the Munitions of War Act, used its powers to compel trade unions (and employers) to accept compulsory arbitration, while 'leaving certificates' were introduced preventing munitions workers from quitting their job to take another without the permission of their employer. A form of industrial conscription was thus being instituted. In return the Ministry, headed by Lloyd George, promised the restoration of pre-war practices once the war was over.[59] However, many craftsmen

> had little faith in the [Government] pledge that no change in practice made during the war shall be allowed to prejudice the position of the workmen or their trade unions in regard to the resumption and maintenance after the war of any rules or customs existing prior to the war.[60]

The other area of conflict, which surfaced in 1915, was over the introduction of women into engineering workplaces. Sceptical that the 'old order' would actually be restored, craftsmen feared that the substitution of unskilled women and men for skilled workers would become permanent, regardless of government promises. They considered that the admission of women into jobs previously done by men would, after the war, make it harder for men to maintain wages at a sufficient level to sustain their position as the primary income earners in their family. This was a significant demarcation between the skilled and unskilled within the working class. It was also bound up with a specific version of what constituted manhood. Increasingly, these attitudes were reinforced by the

---

57 *The Official History of the Ministry of Munitions*, Vol. 1V, Pt. 2, Ch. 11, p. 47.
58 John Maclean, 'The Clyde Unrest', *Vanguard*, November 1915, p. 5, see John Maclean archives, marxists.org accessed 8 January 2017; *The Official History of the Ministry of Munitions*, Vol. 1V, Pt. 2, Ch. 11, p. 47.
59 See *The Official History of the Ministry of Munitions*, Vol. 1, Pt. 1. Ch. 11, p. 37.
60 *The Official History of the Ministry of Munitions*, Vol. 1V, Pt. 2, Ch. 11, p. 47.

dominant societal assumption that men should be the main breadwinners.[61] The ASE called for equal pay for women replacing skilled men, but this was mainly as a way of insuring employers could not drive down the wages of men when they returned to their jobs after the war rather than because of a principle of equity.

Notably, in the summer of 1915 the ASE signed a formal agreement with the National Federation of Women Workers enabling local committees to be established to determine the wages and conditions of female labour. The aim of this united approach was to give these two unions greater power when their demands were placed before local employers. However, few of these local committees got off the ground as the introduction of statutory orders made their work largely redundant.[62]

By the end of November 1915, the National Federation of Women Workers, formed in 1906 and led by Mary Macarthur, was beginning to register significant increases in membership. In Bristol, since the summer, membership had grown by a third to 1,500.[63] Women employed on munitions, now doing the same work as men, perceived the injustice of unequal pay. Furthermore, women were even more disadvantaged than men under the 'leaving certificate' provisions. The fact that they were on less pay than men made the restriction on their liberty to obtain better paid work elsewhere more galling.

Gender changes in employment during the war highlighted the case for equal pay for equal work. However, as Angela Woollacott observes 'neither the government nor employers wanted to raise women's wages to parity with men's, and the men's unions turned a blind eye as long as men's wage rates were not threatened'.[64] Yet the difficulties of fixing wage rates in regard to the introduction of female and male substitute labour could not be ignored. Equal pay for women had been raised within the trade union movement during the nineteenth century, however during the First World War it was being mooted from several perspectives. The former Labour MP, and member of the National Union of Shop Assistants, James Andrew Seddon, emphasised as much in his presidential address to delegates of the Trades Union Congress held at Bristol in September 1915:

---

61 For debates on the male breadwinner see C. Creighton, 'The Rise of the Male Breadwinner Family: A Reappraisal' in *Comparative Studies in Society and History*, Vol. 38, No. 2, April 1996, pp. 310-337; also see J. Bourke, 'Housewifery in Working –Class England 1860-1914' in *Past and Present*, 1994, 143 (1): pp. 167-197.
62 B. Drake, *Women in Trade Unions*, p. 75.
63 *Western Daily Press*, 30 November 1915, p.5.
64 Woollacott, *On Her Their Lives Depend: Munitions Workers in the Great War*, p. 115.

> The question of women's labour, owing to war conditions, is not a passing phase: their introduction into many trades is causing misgiving, and in some cases hostility. This is to be regretted. The only course, to minimise any possible danger, is to insist upon equal pay for equal work. Sex antagonism is the sure way to lower existing standards.[65]

He went on to congratulate the National Union of Railwaymen for securing recognition of equal pay for women from some companies. The Railwaymen's union opened their doors to women for the first time in 1915.

In September, to address the problems emanating from the Munitions of War Act, the Central Munitions Labour Supply Committee established a wages sub-committee. It consisted of Mary Macarthur,[66] General Secretary of the National Federation of Women Workers, James Kaylor, Bristol member of the Executive Council of the ASE, Allan Smith, Secretary to the Engineering Employers' Federation and Glyn West, representing the Supply Departments of the Ministry of Munitions. It worked on the principle that no skilled man should be employed on work that could be done by semi-skilled or unskilled male or female labour. On this basis they drew up recommendations for the payment of women and girls. The main committee agreed to these and, after a meeting with the ASE, accepted further clauses concerning the wages of unskilled men under dilution schemes.[67]

In order to gain the support of the engineering unions, semi-skilled or unskilled workers, who did work *identical* to that previously executed by fully trained and qualified engineers, had to be paid the same time-rates and piece-rates as the men they displaced. The aim was to safeguard the wage-rates of time-served tradesmen.[68] The regulations for the payment of women and girls doing work customarily done by men were set down in a circular (L.2) sent to all the new government-owned factories, and to controlled commercial establishments. The circular stated that a woman employed in the place of a semi-skilled and unskilled man should be paid at his piece-rate but *not* at his time-rate; a loophole that allowed employers to pay women less than the men

---

65 *Western Daily Press*, 7 September 1915, p. 4.
66 Mary Macarthur, a Glaswegian, was described by Glasgow shop stewards, at a meeting addressed by Lloyd George on Christmas morning 1915, as 'the best man o' the lot' (a report of the meeting in *Forward*, the mouthpiece of the local ILP in Glasgow, cited in *The Official History of the Ministry of Munitions*, Vol. 1V, Pt. 4, Appendix X1X). She used her position on the sub-committee to fight 'for an equal rate of pay for equal work and for an adequate living wage for all women workers.' (S. Lewenhak, *Women and Trade Unions* (Ernest Benn, 1977), p.145).
67 *The Official History of the Ministry of Munitions*, Vol. V, Pt. 2, Ch. 11, (HMSO, 1922), p. 11.
68 *The Official History of the Ministry of Munitions*, Vol. V, Pt. 2, Ch. 11, p. 12.

they replaced. However, the L.2 circular, guaranteed to pay women on men's work a minimum wage of £1 per week, and this minimum 'tended to become the standard' rate.[69]

> Women of eighteen years of age and over employed on time work customarily done by men, shall be rated at £1 per week, reckoned on the usual working hours of the district in question for men in Engineering Establishments.[70]

These regulations in circular L.2 were, however, not binding on controlled establishments, of which there were thousands. Consequently, despite agitation for their observance by the National Federation of Women Workers they were not universally adopted.[71]

However, confronted by the prospect of women taking industrial action, and the ASE threating non-cooperation with the Ministry of Munitions dilution policy, the government succumbed. Legislation was subsequently passed enforcing the regulation of women's pay according to the terms specified in L.2. A further circular, L.3, covered the substitution of semi-skilled men for craft workers. In Bristol these men were organised in the main by two unions, the Dock, Wharf, Riverside and General Workers' Union and the Workers' Union.[72] Bristol's aircraft works were thus affected by developments occurring nationally: the State promoting the regulation of pay, the introduction of women, organised labour and pressuring men to join the armed forces.

Craft workers' reaction to the Munitions Act

Opposition to dilution appeared in Bristol's aircraft factories during 1915, around the same time as the Admiralty appointed the Bristol firms, Parnall & Sons, and Brazil Straker & Co., as contractors for the manufacture of aeroplanes and engines for the military. Parnell's operated across several sites. It made propellers at its factory in Mirvert Street, Easton, doped (varnished) the surface of linen covered wings and fuselages at Quakers' Friars, in central Bristol, and assembled aircraft at its premises in Park Row, producing its own design for coastal defence aircraft. By June 1915, Parnall & Sons had begun work on

---

69 Braybon, *Women Workers in the First World War*, p. 54.
70 *The Official History of the Ministry of Munitions*, Vol. V, Pt. 2, Ch. 111, p. 16.
71 *The Official History of the Ministry of Munitions*, Vol. V, Pt. 2, Chapter 11, p. 13.
72 R. Whitfield, 'Trade Unionism in Bristol 1910-1926' in I. Bild (ed.), *Bristol's Other History* (Bristol Broadsides, 1983), p. 81.

an Admiralty contract for Shorts' seaplanes.[73] Brazil Straker assembled Rolls-Royce Hawk and Falcon engines at its factory in Fishponds.[74] Brazil Straker would go on to produce 2,500 Roll-Royce engines by the end of the war.[75]

As orders were placed, these companies turned to the hiring of women and girls mainly for assembly work, but also for working on metal, partly to replace men who had enlisted for military service, and partly to meet the exponential increase in demand for defence and fighter aircraft.[76] This labour shortage problem was compounded by the introduction of a new scheme for raising volunteers, the Derby Scheme, named after Lord Derby who had been appointed in October 1915 charged with boosting voluntary enlistment numbers or securing attestments from men to serve if called up later on. 'Eminently satisfactory' were the words used by the contemporary pro-war newspaper *Bristol and the War* in its assessment of the Derby Scheme in Bristol.[77] While this description most likely overplayed the reality of the number of volunteers coming forward the pressure to import women as substitute labour continued.

During 1915, the demand for dilution fostered tensions in workplaces geared to production necessary for the war. While the leadership of craft unions, such as the ASE, were closely integrated with State policy, serious rank-and-file resistance first erupted in Scotland where a shop steward movement with revolutionary ideas had already emerged out of established workshop organisations on Clydeside, an area in and around Glasgow. Though dominated by shop stewards representing skilled engineering workers, the Clyde Workers Committee recognised that it was politically desirable 'to break down divisions between craftsmen and less skilled workers, to develop an industrial policy which united [their] interests'. [78] Nonetheless, in defiance of the Executive Council of the ASE, the Clyde Workers' Committee demanded a price for its cooperation with the dilution of labour. They insisted that organised labour must have a share in controlling the introduction of dilution on terms that would not be disadvantageous to either unskilled or, in particular, skilled

---

73 C. L. Freeston, 'The British Aircraft Industry' in *Aeronautical Engineering: supplement to "The Aeroplane"* 2 November 1917, p. 1337.
74 Grace's Guide to British Industrial History, gracesguide.co.uk accessed 1 January, 2017: Barnes, 'Bristol and the Aircraft Industry, *Bristol Industrial Archaeological Society Journal* 34, 1972, accessed online 2 January, 2017, www.b-i-a-s.org.uk; J. Penny, *Bristol at Work* (Derby: Breedon Books, 2005) PP. 162-4 and pp.178-9.
75 K. Wixey, *Images of England: Parnall's Aircraft* (Tempus Publishing, 1998) p. 26.
76 Freeston, 'The British Aircraft Industry', p. 1337.
77 *Bristol and the War*, Vol. 11. No. 22, 1 December 1915.
78 J. Hinton, 'Introduction' in Reprints in Labour History No. 1 to J. T. Murphy *The Workers' Committee: An Outline of its Principles and Structure* (first published by the Sheffield Workers' Committee 1917, Pluto Press, 1972) p. 3.

Women & Men working on the construction of Bristol F2B Fighter: Wood Details.

workers.[79] An outraged Lloyd George responded: 'It would be a revolution, and you can't carry through a revolution in the midst of war.'[80]

A core of shop stewards, being indeed on the revolutionary left, was undeterred. On Christmas Morning 1915, at a meeting chaired by Arthur Henderson MP (Labour Party leader and member of the War Cabinet) Lloyd George faced around 3,000 shop stewards and trade union delegates in St. Andrew's Hall Glasgow to argue the case for dilution. A chorus of hissing and booing greeted Henderson, who responded by naming several leading trade unionists who were working with the Coalition government — among them was the Bristol member of the Executive Council of the ASE, James Kaylor, an uncritical supporter of the labour dilution scheme.[81] When Lloyd George rose to speak, militant Glaswegians and shop stewards from across the country interposed singing 'two verses of "The Red Flag"… before [he] could utter a word.' His long appeal for cooperation was frequently interrupted. John Muir got up on a chair to speak on behalf of the Clyde Workers' Committee. Although

---

79 Murphy, *Preparing for Power*, p. 120.
80 Cited in Hinton, *The First Shop Stewards' Movement*, p. 135.
81 *Birmingham Mail*, 27 December 1915, p. 4.

he had been promised a hearing, and despite several attempts to convey his message, Henderson and Lloyd George prevented him from explaining the Workers' Committee's position and the meeting broke up in disarray.[82]

The government were very concerned about the influence of the shop stewards movement spreading beyond Clydeside to other munition centres around the country. It became consumed with halting its progress, especially the bringing together of skilled and unskilled workers and the establishment of links with anti-war individuals and campaigning groups such as the No Conscription Fellowship.[83] In the New Year, the government reacted strongly against the dilution struggle on Clydeside by enforcing the Munitions Act with sackings, imprisonments and deportations. Defeated on the Clyde, the dispersed shop stewards, although closely watched by the Ministry of Munitions intelligence agents, had an impact in other British munitions centres.

Concern about supposed German plots to sabotage transport systems and munitions factories furthered wartime patriotism. A series of local panics occurred such as the alleged attempt to wreck a train on a line near Avonmouth, Bristol, in 1915.[84] Along with the desperate shortage of military equipment these made rebellious workers an easy target. Despite the increasing use of scientific management methods based on premium bonus, rate fixing and the sub-division of labour,[85] commentators were accusing workers of restricting output. Correspondence appeared in the press, under government influence, denouncing slackers and shirkers. A long article in the patriotic magazine, *The Aeroplane,* written by its editor Charles Grey Grey, or C. G. G. as he signed his articles, berated British workmen:

> If, owing to the slackness of the British workman, German troops land in this country because of shortage of guns and ammunition, or because there are not enough aircraft to spot the approach of the invading fleet, and if those German troops commit in English coast towns all the outrages they have committed in Belgium, I say the workmen…are greater criminals than the men who commit the outrages.[86]

82 Taken from a report of the meeting in *Forward*, the mouthpiece of the local ILP in Glasgow, cited in *The Official History of the Ministry of Munitions*, Vol. 1V, Pt. 4, Appendix X1X.
83 S. Rowbotham, *Friends of Alice Wheeldon: The Anti-War Activist Accused of Plotting to Kill Lloyd George* (Pluto, second edition, 2015, first published in 1986), p. 45.
84 *Manchester Evening News*, 2 February 1915, p. 3.
85 See K. Whitson 'Scientific Management Practice in Britain, A History' (unpublished PhD, University of Warwick, 1995) Chapter 2 for an in-depth study of scientific management practice before and during the First World War.
86 *The Aeroplane*, 10 November 1915, p. 569.

Two weeks later, 24 November 1915, *The Aeroplane* published a letter of support for C.G.G. from the aristocratic civil engineer, Lord Headley, the fifth Baron of Headley, suggesting imprisonment for the labour agitators who he claimed were responsible for promoting the practice of output restriction and inciting labour unrest:

> The only place for people who, with deliberate intent to diminish the country's turnout of munitions, spread poisonous doctrines amongst the workmen, is prison, and after the war trial for treason or treason felony.[87]

And in the 12 January 1916 edition of *The Aeroplane*, C.G.G. argued that slackers and shirkers should be drafted into the army and that 'out-and out-conscription' must be introduced 'tempered only by intelligently and justly administered tribunals of appeal.'[88]

The vilification of shop stewards and their supporters heightened the existing tensions, particularly those created from the divisions between the rank-and-file engineers and the official union leadership's support of the government on the dilution question. These divisions had not been eradicated by the defeat imposed by the government on the Clyde shop steward movement. Clyde deportees, such as Arthur MacManus, took their experiences and forms of organising to other munitions centres south of the Scottish boarder. Despite the influence of Arthur Henderson's ally, James Kaylor, as the representative of the city's engineers on the Executive Council of the ASE, signs of dissension appeared in Bristol.

Both dilution and the threatened introduction of military conscription continued to be the causes of conflict in the New Year. The difficulties faced by Lloyd George, in 1916, demanded political astuteness if he was to fulfill the requirements of the military and carry public opinion. He used all his skill and determination to push ahead with what he saw as the necessity of dilution as the only means of increasing the output of munitions and, after the failure of the Derby Scheme, the inevitability of conscription.[89]

Though the government formed close links with employers, Lloyd George recognised the need to bow to pressure from the ASE and the National Federation of Women Workers in order to achieve his objectives. Consequently, the Munitions of War (Amendment) Act of 27 January 1916 instructed that in government controlled establishments 'the Minister of Munitions shall have

---

87 *The Aeroplane*, 24 November 1915, p. 640.
88 *The Aeroplane*, 12 January 1916, p. 92.
89 C. Wrigley, *Lloyd George* (Blackwell, 1992), p. 74.

power by Order to give direction as to the rate of wages of the female workers so employed'.[90] This Order also applied to the remuneration of semi-skilled and unskilled men who were doing work customarily carried out by skilled men. In February 1916 the regulatory orders (L.2 & L.3) gained Royal Assent and became legal and mandatory.[91]

However, these orders were not comprehensive. This legislation enforcing, at the very least, a minimum wage for women did not cover female substitutes employed on woodwork and sheet metal work for aircraft. The Ministry of Munitions referred the position of these women to a Special Tribunal. On 17 August 1916, after a series of conferences between members of this Tribunal with aeroplane manufacturers and trade unions, a set of recommendations was issued for the payment of women woodworkers in the industry. On implementation, however, these proposals carried with them the same problems as those experienced under the Circular L.2, particularly where women were only doing part of the work previously carried out by skilled men. In some cases this was because they were often excluded from learning and executing the full range of craftsmen's work. The National Amalgamated Furnishing Trades Association, to which most of the women woodworkers in the aircraft industry belonged to, vehemently opposed the employment of women on woodcutting machinery, ostensibly because of the dangerous nature of this skilled work.

Bristol workers' response to dilution and conscription

Although attention has been paid to shop floor resistance to dilution in major armaments' works in Scotland, the North, Midlands and London, Bristol has been overlooked. The first indications of hostility appeared in Bristol early in January 1916. Just a few weeks before the passing of the 1916 Munitions of War (Amendment) Bill and the Military Service Bill, the Bristol Sheet Metal Workers' Union, whose members worked mainly on 'fuel and oil tanks, engine cowlings and wheel spats,'[92] had taken action against one of its members for, allegedly, training women on skilled men's work. On 10 January 1916, the Metal Workers' Union informed a forty-three year-old worker, Albert Pole 'that in the interests of this Society he should leave his present employment [at Parnall & Sons, Bristol], as it was understood this member was learning girls at Parnals (*sic*), part of our work.'[93] This demonstrates the extent of union power

---

90 *The Official History of the Ministry of Munitions*, Vol. V, Pt. 2, Chapter 11, p. 15.
91 Drake, *Women in Trade Unions*, p. 79.
92 Brake, *Men of Good Character*, p. 148.
93 Minutes of the General Meeting of the Bristol Operative Tin and Iron Plate, Sheet Metal Workers and Braziers, 10 January 1916.

during the war. Holding a trade union card was an important credential, as unions representing skilled workers would, with few exceptions, only accept time-served workers into membership.

Opposition to dilution, however, was not confined to the workers. Many employers strenuously objected[94] because they faced shortages of skilled labour and also feared provoking unrest. Nevertheless, acceleration in the pace of recruiting women for munitions work, together with improvements in production efficiency based on premium bonus, rate fixing and the sub-division of labour, persisted, even while the Bill for the introduction of conscription was being debated in the House of Commons.

Meanwhile resistance to conscription was attracting some public support. On 17 January 1916, under the auspices of the Bristol Trades Council and the Labour Representation Committee, a well-attended protest meeting, held at Kingsley Hall, Old Market, overwhelmingly passed the following resolution proposed by Bristol Councillor, and ASE member, Walter Ayles:

> That in view of the renewed activity of the Conscriptionists inside and outside the Cabinet, and the introduction of a Conscription measure in the House of Commons, this meeting reaffirms its unalterable opposition to conscription. It has closely followed the agitation for conscription from the outbreak of the war, and holds that the purpose is not for securing any national advantage, but has been pursued for the purpose of smashing the trade union movement and undermining the power of the British democracy. We repudiate the trick of dividing the nation's manhood into two parts, namely, married and unmarried, and pledge ourselves to resist the introduction of this Prussian principle to conscript human life in any form whatsoever.[95]

Ayles' socialist politics had been formed by a left which stressed the importance of ethics. Although he and his fellow opponents to the war were in a minority their influence would become wider after the war when anti-militarism gained more support. In 1923 Ayles was to be elected as a Labour MP for Bristol North.

However, after the passing of the Military Service Bill on 27 January 1916, the conscription of single men between the ages of 18 and 41 was ordered,

---

94 C. More, *Skill and the English Working Class, 1870-1914* (Croom Helm, 1980), pp. 28-29.
95 *Western Daily Press*, 17 January 1916, p. 7. For an account of Walter Ayles' activity as a conscientious objector see C. Thomas *Slaughter No Remedy: The Life and Times of Walter Ayles, Bristol Conscientious Objector* (Bristol Radical History Group, Pamphlet 36, 2016). See also J. Hannam, *Bristol Independent Labour Party: Men, Women and the Opposition to War* (Bristol Radical History Group, Pamphlet 31, 2014).

Men working on the construction of Bristol F2B Fighter: Sheet Metal Shop.

which significantly increased the pressure for the substitution of skilled male labour by women in Bristol's aircraft factories. By 20 March, concerns about how output could be increased triggered a change of attitude at the B&CAC. The managing director of the company, G. Stanley White, advised his executives that 'women labour should be introduced into the Works, and that we should continue to employ women labour after the War.'[96] 1916 was to see the release to the military those men who had previously been considered indispensable.

The enactment of Military Service Act in March led to an increase in the numbers of people prepared to protest. In March 1916, Sir John Simon, who had resigned from the Asquith Liberal government in order to take a stand against conscription, continued to oppose it arguing that: 'Any chance of compulsion becoming popular had been destroyed by the ineptitude of its advocates and administrators.'[97] Moreover, a series of rallies against the war were held over April and May. A twenty thousand strong Trafalgar Square stop-the war demonstration on 8 April 1916 was followed by an Easter Sunday anti-conscription gathering of over two hundred thousand at the same venue,

96 The British and Colonial Aeroplane Company Minute Book, No. 1, 29 March 1916.
97 *Western Daily Press*, 17 March 1916, p. 5.

which was 'broken up by patriot violence'.⁹⁸ Regardless of the hostility, anti-conscription conferences and anti-war demonstrations were held across the country, including one in Bristol on 7 May attended by several hundred people on the top of Clifton Downs.⁹⁹

In May the government proposed that conscription should be extended to include married men. In Parliament, Simon was again the main voice in opposition. However, he could only muster support from twenty-seven Liberal MPs and ten from Labour.¹⁰⁰ Even though overt opposition to the war was confined to a minority, a diffuse undercurrent of disillusionment with the war nonetheless concerned the authorities. As Lloyd George recorded in his memoirs, '[r]eal Parliamentary opinion can rarely be gathered from a perusal of division lists. There were sinister grumblings in the corridors and tea-rooms' against the activities of the Ministry of Munitions.¹⁰¹

In Bristol, a coincidental consequence of the introduction of conscription for B&CAC employees was the impact it had on the implementation of the 1915 promise of a productivity bonus scheme at the end of the war. Clearly the guarantee of continuous, unbroken employment was now thrown into question. Representations made by the workers to management petitioned 'that the output bonus of £20 per machine should be paid immediately instead of being left until the end of the war.' Clearly workers held strong feelings on this question, which came up for discussion at the October 1916 Directors' Meeting. The decision was taken at this meeting that in future this output bonus would be paid quarterly in the form of War Savings Certificates. In order for White to convey this ruling to the workforce the meeting was adjourned; first, he addressed a mass meeting of employees at the Filton Works informing them of the directors' decision and then he travelled to the Brislington Tramway Works to relay the same verdict.¹⁰² There is no record of the workers' response to this ruling but it is likely they would have derived satisfaction from the knowledge that they had forced the company to make a significant concession. The B&CAC could not ignore the voices of the ASE, whose membership in Bristol had almost doubled between 1913 and the end of 1916,¹⁰³ in combination

---

98 B. Millman, *Managing Domestic Dissent in First World War Britain* (Frank Cass, 2000) pp. 84-5. This event was not reported in the *Daily Herald*, or elsewhere, because as this paper stated: 'Our readers will understand that at present we are unable to comment on the latest phase of the Conscriptionist movement. The latest Press regulations issued under the Defence of the Realm Act are so stringent and so widely drawn that it is almost impossible to say what is legal or illegal', (*Daily Herald*, 29 April 1916).
99 *Western Daily Press*, 8 May 1916. Hannam, *Bristol Independent Labour Party*, p. 32.
100 D. Lloyd George, *War Memoirs of David Lloyd George*, Vol. 11 (Odhams Press, 1938), p. 439.
101 Lloyd George, *War Memoirs of David Lloyd George*, Vol. 11, p. 414.
102 The British and Colonial Aeroplane Company Minute Book, No. 1, 14 October 1916.
103 Whitfield, 'Trade Unionism in Bristol 1910-1926', p. 81.

with the Dock, Wharf, Riverside and General Workers' Union, as well as the Workers' Union and the National Federation of Women Workers. The ad hoc coalition of diverse unions presented a formidable organised force.

Conscription, the labour supply and the increase in demand for women workers

In 1916, the National Advisory Committee on War Output to the Ministry of Munitions, of which James Kaylor, Bristol ASE, was a member, delegated some of its functions to local Labour Advisory Committees in order to draw on their local knowledge on questions of dilution and the release of men for military service. This was also an attempt to popularize its role. Each local committee consisted of seven members elected from those trade unions whose members worked for companies producing munitions of war. Their main function was 'to collect information on labour difficulties…and to facilitate in every way the output of war material,' but their powers were limited to reporting difficulties and grievances to the National Advisory Committee.[104] One such Committee was operational in the West Country covering Gloucester, Somerset, Devon and Cornwall. The Committee comprised the chairman A. Wakeham (ASE); J. D. Shiner (Amalgamated Society of Ironfounders); J. D. Ashmore (Society of Steam Engine Makers); W. Jackson (Dock, Wharf, Riverside and General Workers' Union); L. Biggs (U.K. Smiths and Strikers Union); P. Walsh (Society of Boiler Makers) and G. Piper, (United Pattern Makers Association).[105]

One of its roles was to advise on appeals in cases when local employers disputed decisions ordering the release of men, regarded as 'insufficiently employed', for military duty. This was significant as 'the exemption of munition workers from military service depended primarily on their employers.'[106] Some employers were classifying 'men who were paid unskilled rates as 'skilled' in order to keep them.'[107] Clearly this was an extremely sensitive issue and even though the Committee consisted solely of trade union official representatives the fact that in Bristol it did not sit in public, and it had connections with the military recruiting office, prompted the Liberal Alderman, John Swaish, to question its integrity.[108]

Another example of local involvement was the formation of a Bristol branch

---

104 *The Official History of the Ministry of Munitions*, Vol. 1, Pt. 2, Ch. 1V, p. 97.
105 *Western Daily Press*, 4 March 1916, p. 9.
106 *The Official History of the Ministry of Munitions*, Vol. 1V, Pt. 1, Ch. 1V, p. 55 and Vol. V1, Pt. 1, Ch. 1, p. 2.
107 R. Aris, *Trade Unions and the Management of Industrial Conflict* (Macmillan Press, 1998), p. 118.
108 *Western Daily Press*, 30 June 1916, p. 3.

of the Committee on Women's War Employment which held the important job of collecting information 'as to the labour requirements of employers in their area,' and 'the organisation of a supply of women workers'. Their remit extended beyond the workplaces to the social needs of women workers including 'the making of arrangements for housing accommodation for women brought in from other districts, and the initiation of schemes for the welfare of women employed in their area.'[109]

The local liberal newspaper in Bristol, the *Western Daily Press*, rallied to the patriotic cause with an emollient article which sought to reassure male skilled workers. The dilutee is presented as a serving soldier's sister sacrificing herself like 'Tommy' (the name commonly ascribed to the British private soldier). She is depicted in smoothly, patronising terms.

> It is done under greater strain and at a greater cost than many people realise. She has not grown into the work as Tommy has done; it is all new and correspondingly hard. She meets the new technical difficulties every day; she makes mistakes which exasperate her over worked employer, who is often slow to make allowances for the lack of experience and training which causes them… She is often unaccustomed to steady hours of work or confinement in a close atmosphere, and suffers physically from these causes. To take up work she may have left her home in some country town or village and come into the city, where possibly she finds herself friendless…Too often her only home is some lodging far from comfortable, very far from cheerful…[110]

A more nuanced account of the pros and cons of war work is preserved in the Imperial War Museum's sound Library of a Somerset farmer's daughter, Elsie Hilliar (her married name), who had worked first on munitions in Coventry and then on aircraft wings at the Filton Works of the B&CAC in Bristol.[111]

In 1915, Elsie Hilliar, weary from rising at 4 am every morning to milk the cows, ran away from her father's farm, in Welton near Radstock, with her sister to do what she felt was 'real work' at the Coventry shell filling and fuse factory managed by White & Poppe Ltd. for the Ministry of Munitions. She had passed a qualifying exam to work on munitions. Her account reveals how her life was

109 *Western Daily Press*, 1 July 1916, p. 5.
110 *Western Daily Press*, 13 November 1916, p. 7.
111 Elsie Hilliar (interview by Chris Howell, 1982), Imperial War Museum Sound Archive, Cat. No. 6682, Reel 1. It is hard to find recordings of women who had worked in aircraft factories during the First World War. Hilliar was the only one I could find who had been employed on aeroplane production in Bristol.

transformed. She would have worked long hours with an army of women under the watchful eye of an overlooker but no doubt looked forward to pay day and perhaps the opportunity to spend a little on a weekend's entertainment. If pay for a woman like herself was higher, so were the prices of basic goods. Never having experienced hunger on the family farm she remembered it while at Coventry that: 'cor, didn't we starve...All's we had was bread with a scraping of whatever we could find. Mustard!'[112]

During 1916, however, White & Poppe reduced the output of fuses to make way for wing production.[113] Elsie Hilliar's interview records how the introduction of Scottish women into the Coventry works had put her and some of her friends out of work. She recalled having no work and needing to do something about that. By chance she knew a man at the Bristol Labour Exchange and he found work for her and two friends at the B&CAC Filton Works. So she was put on night work covering wings with linen. During the week she lodged at a house in Doone Road, Horfield, around two miles from the Filton Works, but at weekends after finishing her Friday night shift she cycled to her parents home in Welton, twenty-one miles away, and then on Monday cycled back in time for the nightshift.[114]

Although Hilliar saw her war work as an improvement on her life at the farm, women workers like her faced resentment and real dangers. In the workplace negative attitudes towards substitute women could still be found. Notwithstanding the low opinion some employers and skilled men had for substitute workers, women soon predominated in areas of work such as sewing linen onto wing frames, and the dangerous task of doping the linen covered wings, to make them air tight and waterproof. This process involved applying a highly toxic cellulose compound containing tetra-chloride. Long-term exposure to this substance could be fatal.[115] A woman working on varnishing aeroplane wings at the Royal Aircraft Factory in Farnborough died from the inhalation of tetra-chloride vapour. The inquest returned a verdict of accidental death, although as a consequence of this 'accident' the factory inspector recommended the introduction of improved methods of ventilation.[116] In 1916, twenty-six cases of tetrachlorethane poisoning were reported, all of which concerned

112 Elsie Hilliar (interview by Chris Howell, 1982); Also quoted in Chris Howell, *No Thankful Village: The Impact of the Great War on a Group of Somerset Villages- a microcosm* (Bath: Fickle Hill, 2002), p. 104.
113 L. A. Batchelor, 'A Great munitions centre: Coventry's armaments and munitions industry 1914-1918.' (Unpublished MScR Thesis. Coventry: Coventry University, 2008), p. 104.
114 Elsie Hilliar (interview by Chris Howell, 1982).
115 A. Hamilton, 'Dope Poisoning in the Manufacture of Airplane Wings', *Monthly Review of the U.S, Bureau of Labor Statistics*, Vol. 5, No. 4, October 1917, pp. 18-25.
116 *Coventry Evening Telegraph*, 22 June 1915, p.2.

aircraft workers using dope, and five of these were fatal.[117] At the Brislington works of the B&CAC, and Parnall's Aircraft works, in Bristol, women working at doping aircraft wings reported experiencing giddiness and headaches as a result of exposure to toxic fumes.[118]

The variety of women's work in aircraft manufacture included 'assembly, finishing machine-made struts, polishing propellers' and many more operations.[119] Most of the women were either organised in the National Amalgamated Furnishing Trades Association, the National Federation of Women Workers, or other non-craft unions admitting women to membership. The Amalgamated Society of Carpenters and Joiners were adverse to women joining its union and had serious misgivings about dilution.[120] At the same time, the male woodworking unions were struggling to secure for their members the same recognition of their skill, from the Ministry of Munitions, as that given to the engineers and metal trades. Unions representing aircraft woodworkers had lobbied hard to obtain uniform rates of pay for a wide range of woodcraft skills — joiners, cabinet-makers, coachbuilders, organbuilders — 'equal to that of the best paid trade from which workmen had been recruited' for war work without success.[121] The unions' determined opposition to the imposition of piece-work in the woodworking industries added to their list of grievances.[122]

On 7 December 1916, Lloyd George replaced Herbert Asquith as Prime Minister to lead a coalition government. In Bristol a policy of combing out men for military service, who had previously been given exemption by the local Labour Advisory Committee, then began.[123] In April 1917, the *Western Daily Press* reported that this policy had been successful,[124] though some still questioned whether the process had been 'as searching as it ought to have been'.[125] The focus of such complaints switched from the local Labour Advisory Committee to local Munitions Tribunals, as they were the only bodies that had the power to deliver exemption from military service.[126] This could lead to a labyrinthine process. One example is that of Pritchett & Gold

117 A. Keith Mant, M.D. 'Acute Tetrachlorethane Poisoning' in *The British Medical Journal*, Vol.1, 21 March 1953, p. 655.
118 Bristol Museums Galleries Archives: Industrial and Maritime History, object number: J4747 (photographs)'Cabinet Shop; Sheet Metal Workers; Wood Details; Doping Wings; Final Erecting; Assembling Planes'.
119 *The Official History of the Ministry of Munitions*, Vol. V, Pt. 2, Ch. VII, p. 90.
120 *The Official History of the Ministry of Munitions*, Vol. V, Pt. 2, Ch. VII, p. 89.
121 *The Official History of the Ministry of Munitions*, Vol. V, Pt. 1, Ch. IV, p. 96.
122 *The Official History of the Ministry of Munitions*, Vol. V, Pt. 2, Ch. VII, p. 89.
123 *Western Daily Press*, 9 December 1916 *Western Daily Press*, 10 January 1917, p. 7.
124 *Western Daily Press*, 30 April 1917, p. 5.
125 *Western Daily Press*, 16 May 1917, p. 4.
126 *The Official History of the Ministry of Munitions*, Vol. IV, Pt. 3, Ch. III, p. 85.

Women working on the construction of Bristol F2B Fighter: Doping Wings.

Electric Power Storage Company, based in Dagenham, who employed Bristol resident, Joseph Entwistle, as a Traveller and Battery Inspector. Early in 1917, the company appealed, through the agency of a Munitions Tribunal and the Labour Enlistments Complaints Committee, against Entwistle being called up for military service. The appeal was successful and an exemption certificate was issued from the Bristol Recruiting Office but with a caveat. He had been attested fit for Garrison Service in June 1916, despite being rejected as unfit when he had volunteered for the army in November 1915, and therefore could still be drafted at short notice by the Recruiting Officer (see Entwistle's call-up cards, certificates and letter in the appendix). He remained in limbo until the war ended when, on 28 November 1918, he was discharged from the 2nd (Bristol) Volunteer Force, Gloucester Regiment having never been called up.[127]

A result of the stepping-up of the combing out policy was yet a further increase in demand for women to take up many of the positions vacated by men conscripted to serve in the armed forces. Elsie Hilliar and her two friends were swept up in this broader recruitment drive. On 17 May 1917, the local

127 See Joseph Entwhistle (Entwistle) papers, Bristol Archives, Ref. 37267/1-10.

press carried an advertisement calling for women and girls to train on aero-engine work, most likely at Brazil, Straker & Co.: 'Respectable girls required to train for the Assembling of Aircraft Engines, etc. Wages 25s per week during training'. Applicants were asked to apply to the Labour Exchange, Victoria Street, Bristol.[128] The offer of being paid by the government for learning a trade, along with the expectation that after several weeks training they would receive a comparable, if not higher wage than they could obtain from other employers, would have appeared attractive to some young women, regardless of the gender pay disparity. This redoubling of the effort to recruit women workers in Bristol, and elsewhere, coincided with heightening labour unrest in engineering workshops.

Elsie Hilliar's interview gives a hint of this when she describes witnessing sabotage at the Filton Works of the British and Colonial Aeroplane Company: 'After one plane was rolled out it came to light that 'somebody [had] put a penknife through [the fabric] of the wings'. She said it was done 'for fun, that's what used to happen.'[129] Fun or not this would have been regarded severely by the Ministry of Munitions Labour Intelligence Division (known as the MMLI) if they had heard of it. By 1917, MMLI had begun to investigate any actions 'likely to lead to munitions not being produced in sufficient quantity, either from the activities of German agents, or from extremists among the British workmen (sic) and persons working in the munition factories.' Members of this special unit (outsourced and private agents) were paid to report intelligence on strikes, labour unrest generally and sabotage.[130]

Labour Unrest: Class and Gender Divisions

Several of Bristol's aircraft factories, like the B&CAC, were unionised. Unionisation did not imply that aircraft workers were necessarily politically militant but in instances when they felt threatened, or treated unfairly, they could react strongly. Grievances arose because the 1916 Military Service Act had introduced conscription for the first time while the Munitions of War Act had enhanced the power of employers, 'who expected and obtained substantial profits in return for their contribution to the war economy.'[131] Wartime measures, however, fuelled discontent across the country amongst munitions

---

128 *Western Daily Press*, 17 May 1917, p. 2.
129 Elsie Hilliar (interview by Chris Howell, 1982); Some but not all of the material in this paragraph also appears in Chris Howell, *No Thankful Village*, pp.104-5.
130 See N. Hiley, ' Counter-espionage and security in Great Britain during the First World War', *English Historical Review*, 1986, p. 653.
131 G. Hardach, *The First World War 1914-1918* (Pelican, 1987, first published in English by Allen Lane, 1977), p. 193.

workers, who were facing high inflation, food shortages and fatigue from grafting long hours in poor working conditions. The handling of these labour grievances, recorded Lloyd George, 'gave Government constant anxiety.'[132]

However, at the end of March the government pushed through an amendment to the Munitions Act to extend dilution beyond the State controlled factories to commercial work, thus enabling semi/unskilled workers, particularly women, to perform work previously carried out by craftsmen. This enabled the transfer of skilled labour to munitions work. This decree broke previous government pledges given to the engineers, testing their patience to the limit. [133]

This simmering discontent boiled over in May 1917 as 200,000 engineering workers nationwide withdrew their labour, including those employed in Bristol's aircraft manufacturing firms.[134] A highly contentious government decree triggered the strike. In opposition to its Executive Council, rank-and-file members of the ASE struck work in protest at the government's decision to abolish the Trade Card scheme (whereby skilled craftsmen were exempted from military service). The government's intention was to release larger numbers of men for conscription into the armed forces. Defence of the boundaries demarcating craft skills became paramount.

The dilution question was complicated because in practice male craft workers, as part of their job, did in fact do varying degrees of routine work. The utilisation of machine tools and the introduction of automation had already begun to alter old practices. Now some of these tasks were being allocated to women. Bristol workers' fears were heightened by the appearance in the local press of advertisements calling for hundreds of girls to train at the Instructional Aircraft Factory at Gloucester Road, in north Bristol, as aircraft workers.[135] The range of jobs included practical engineering, such as the operation of semi-automatic and fully automatic lathes, and work on all classes of woodworking machinery and pattern makers' tools and appliances.[136] This training school opened on 26 June 1917, operating on the site, loaned by Bristol Corporation, of a partly completed swimming baths converted for its new purpose. The school's target was to train around 250 young women, 'of the right stamp, of good physique, and with sufficient education', every six or seven weeks.[137]

While male craft workers' concerns focused on the women, direct

132 Lloyd George, *War Memoirs of David Lloyd George*, Vol. 11, p. 1145.
133 See Hinton, *The First Shop Stewards' Movement*, pp. 196-98.
134 The British and Colonial Aeroplane Company Minute Book, No. 1, 26 June 1917.
135 See *Western Daily Press*, 21 May 1917, p. 2.
136 See for example the Ministry of Munitions of War *Loughborough Technical Institute Instructional Factory*, January 1918, lboro-history-heritage.org.uk accessed 12 January 2017.
137 *Western Daily Press*, 27 June 1917, p. 4.

substitution was less common than was imagined. In fact between April and June 1917 the average number of women and girls on skilled men's work in Bristol was eighty-seven. In contrast the number replacing semi-skilled or unskilled men was 4,950 and on woodwork for aircraft, 200. [138] Nevertheless, in 1917 a cluster of resentments coalesced. Men, whose craft endowed them with a sense of a legitimate privilege, both as men and as workers, defended craft status. In a more abstract sense they were inclined to consider that the direct encroachment of the State into their working lives was a breach of their individual rights. This had a tangible aspect, for the loss of male craft status opened the possibility of being liable to call-up for military service.

In Bristol, where the workforce had shown few outward signs of militancy, a shift occurs in May 1917. Resentment against the stepping up of dilution and the extension of conscription resulted in sympathy for the unofficial strikes sweeping through the country. The strike wave spread to Bristol when, on 12 May, ASE members in the city downed tools. [139] The Government, faced with increasing national unrest, reacted forcibly. On 18 May, seven strike leaders, S. Burgess, W. Hill from Sheffield; T. W. Dingley, N. Cassidy from Coventry; G. Peet, P. H. Kealey from Manchester; A. MacManus from Liverpool,[140] were arrested while attending a conference of shop stewards, at Walworth in London. On the following morning W. F. Watson from London shared the same destiny. They were charged with inciting the recent strikes in the engineering trade, and remanded in custody. The remaining delegation of shop stewards invited the executive council of the ASE to their conference to debate how best to petition for the release of their arrested comrades. The meeting authorised the executive council to negotiate with the Ministry to obtain the release of the arrested men and secure assurances of no further detentions, and no victimisation. It was agreed that the stewards should return to their districts and advise their members to go back to work. According to the official account, the eight detained men signed a statement that they would adhere to the agreement reached by the executive council of the ASE and the Ministry of Munitions.[141]

The engineers did return to work on the recommendation of their stewards, though some areas held out for a few days and Bristol was among them. Bristol

---

138 *The Official History of the Ministry of Munitions*, Vol. V, Pt. 2, Ch. VIII, p. 106.
139 *The Official History of the Ministry of Munitions*, Vol. VI, Pt. 1, Ch. V (HMSO, 1922) p. 112 ftn. 3.
140 See S. Rowbotham, *Friends of Alice Wheeldon* (2nd edition, Pluto Press, 2015) for MacManus's role in assisting conscientious objectors on the run from the police.
141 *The Official History of the Ministry of Munitions*, Vol. VI, Pt. 1, Ch. V, pp. 114-19. Thomas Bell, in his book *Pioneering Days* (Lawrence & Wishart, 1941, p. 125) contends that his friend Arthur MacManus refused to sign the agreement made between the ASE and the Ministry of Munitions and was released nonetheless.

men did not resume work again until 24 May. According to newspaper reports this was with the proviso that they would walk out again 'if a case of alleged victimization in which one [local Bristol] man is concerned is not satisfactorily settled.'[142] This brought to an end the mass strikes that had taken place across the country. Nonetheless, as shop steward committees in large engineering works spread to more towns and cities, including Bristol, rank-and-file sporadic resistance was acquiring a national presence.[143] A new form of activated trade unionism from the base began to emerge. This was exactly what Lloyd George's government, backed by Labour MPs, Arthur Henderson, John Hodge and George Barnes as members of the War Cabinet, had dreaded. Henceforth they could not simply make deals with the official union leadership. Yet the need to secure the manufacture of munitions was imperative. 1,500,000 working days had been lost in the May 1917 dispute, involving around 200,000 men. The government, therefore, sought to compromise. They offered to end the leaving certificate scheme providing dilution was accepted on commercial work. This proved insufficient. Following a ballot of the Engineering Union's membership in July 1917 which, by a large majority rejected the introduction of dilution on commercial work, the government abolished the leaving certificate scheme *and* dropped the proposed dilution clause in the reintroduction of the 1917 Munitions of War Bill that gained Royal Assent on 21 August 1917.[144] The State was thus operating a policy of appeasing male skilled workers while combing out the rest of the workforce for military service. This accentuated antagonisms between workers. A caustic parody, attributed to a soldier, circulating around Sheffield during the May 1917 strike, provides an illustration of the bitter feeling that prevailed among some soldiers and unskilled workers:

> Don't send me in the Army, George,
> I'm in the ASE
> Take all the bloody labourers,
> But for God's sake don't take me.
> You want me for a soldier?
> Well, that can never be —
> A man of my ability, And in the ASE![145]

---

142 *The Times*, 22 May 1917, p.7; see also *Western Daily Press*, 22 May 1917, p. 5.
143 J. T. Murphy, 'The Shop Stewards and Workers' Committee Movement' in A. Gleason, *What The Workers Want: A Study of British Labour* (New York: Harcourt, Brace and Howe, 1920), p. 191.
144 *The Official History of the Ministry of Munitions*, Vol. V1, Pt. 1, Ch. V, pp. 92, 119-20.
145 Cited in Hinton, *The First Shop Stewards' Movement*, p. 210.

Aircraft factories, including those in Bristol,[146] could not produce planes fast enough to meet demand at this time, leading Dr. Christopher Addison, Lloyd George's replacement as the Minister of Munitions,[147] to put out an appeal for more women workers to undergo training so that they could replace semi-skilled and skilled men conscripted for military service.[148] It was now claimed that women required only two months training to enable them to carry out highly specialised aircraft work.[149] Concern about the welfare of women war workers, many of whom had been drafted from outside Bristol, had resulted in hostels being established. But these were unable to cater for the demand.[150] By November 1916, the Bristol women's hostel in Whiteladies Road had become overcrowded, triggering a plea to the general public to raise funds for the purchase and conversion of the adjoining premises.[151] This took a year to achieve. At the end of October 1917, H.R.H Princess Louise opened the enlarged hostel to accommodate some of the numerous young women who had come to Bristol to engage in war work.[152] That month, the dilution percentage of women and boys under eighteen years in aircraft firms stood at 37.5 per cent. By December this figure had increased to 40.1 per cent in aircraft construction, 43.1 per cent in aero-engine firms and 50.5 per cent in aircraft accessories.[153]

The arrival of this new labour force presented an awkward dilemma for those craft unions which continued to refuse entry to women and unskilled men into their organisations. Significant numbers of women and unskilled men were demonstrating that they were capable of doing many of the tasks previously carried out by the craft elite and thus threatening their monopoly over engineering and woodwork. Gender prejudices were pervasive. Some employers were as unenthusiastic about the arrival of women munitions as the skilled workers. A contentious notice placed on the factory noticeboard of the Bristol Engineering firm, Strachan & Henshaw Ltd., accused women on war work of not working hard enough saying, 'are they by slacking going to allow their husbands and others to be slaughtered for want of shell.'[154] However, attitudes were neither uniform nor static and wartime conditions effected changes.

---

146 D. J. James, *The Bristol Aeroplane Company* (Tempus Publishing, 2001), p. 43.
147 In December 1916, Lloyd George, on becoming Prime Minister, appointed Dr. Christopher Addison as Minister of Munitions.
148 *Western Daily Press*, 12 July 1917.
149 *Hendon & Finchley Times*, 17 August 1917.
150 *The Official History of the Ministry of Munitions*, Vol. V, Pt. 5, Ch. 111, p. 28 ftn. 2.
151 *Western Daily Press*, 18 November 1916, p. 2.
152 *Western Daily Press*, 30 October 1917, p. 5.
153 *The Official History of the Ministry of Munitions*, Vol. X1, Pt. 1, Ch. 111, p. 85.
154 *Bristol Times & Mirror*, 17 May 1917.

Labour Relations: the State, employers, trade unions and workers' representatives

The May 1917 strike supported by aircraft workers in Bristol led to the strengthening of organised labour in the B&CAC. There are indications too of a growing political awareness accompanying the shop steward movement. Several individual links can be traced between shopfloor activism and subsequent Labour Party engagement. At Filton, the shop stewards established a Works Committee. Its secretary was L. T. Taylor, who later, in the Coupon Election of December 1918, became involved in the team supporting the Bristol South Labour Candidate, and ex secretary of the Bristol Trades Council, T. C. Lewis.[155] T. J. Scottow, who later went on to become active in the North Wales Independent Labour Party Federation, chaired the Works Committee meetings.[156]

Moreover, a shift in assumptions about labour relations in Bristol's key industries can also be detected in the process of planning the adjustment from war to peace. In July 1917, a report on how to improve industrial relations and plan towards industrial reconstruction was published. It documents the findings of a joint conference, held the preceding February, between Bristol employers and local trade union leaders. The participants of the February gathering were present in a personal capacity and not as representatives of their organisations. The anti-war, Quaker and Liberal MP, Arnold Rowntree, and chairman of the Bristol Munitions Court, barrister Ernest Handel Cossham Wethered, convened the all male meeting. The Somerset born, and former Bristol carter, Ernest Bevin, National Organiser of the Dock Wharf Riverside and General Workers' Union, was among those present. The report recorded that the key matters that needed addressing were:

(a) The question of the status of the operative.
(b) Mutual ignorance of each other's point of view on the part of Capital and Management and Labour.
(c) The claim of the operatives for a greater share of the wealth produced in industry.
(d) The fear of unemployment.
(e) The reluctance of some employers to recognise trade unions.
(f) The dissociation of the operatives from any responsibility in the control or conduct of industry.[157]

---

155 *Western Daily Press*, 10 December 1918.
156 Letter of condolence from T.J. Scottow to Katharine Bruce Glasier on the death of her husband John Bruce Glasier on behalf of the North Wales Independent Labour Party Federation, 1920, Glasier Papers, University of Liverpool.
157 *Western Daily Press*, 14 July 1917, p.5.

Strachan & Henshaw, Women Munitions Workers 1918.

The summary of their discussions, though phrased in carefully inclusive and consensual terms, contained some startling and potentially radical elements. These wide-ranging issues however were left hanging. When the Commission of Inquiry into industrial unrest in Britain was published in October 1917, it simply states that 'in the Bristol area, relations between certain of the employers and certain of the unions are enlightened and progressive.'[158] Ostensibly, relations between B&CAC and the workers' committee did appear to be operating cooperatively at this time. A series of talks between the two parties, during September and October 1917, just a few weeks after the formation of the Ministry of Reconstruction, concluded with a settlement over changes to the output bonus. The system of giving out war saving certificates quarterly was dropped and replaced by a war bonus of ten per cent on wages and overtime paid weekly in cash (backdated to 14 September 1917) until the end of the war.[159]

This favourable result, however, can be accounted for because of the newly established unity between craft and non-craft labour under the umbrella of a works' committee, which was led by politically conscious shop stewards rather than any consensus between the management and its workforce. Moreover,

---

158 *Commission of Inquiry into Industrial Unrest*, No. 6 Division. Report of the Commissioners for the Southwest area, 1917/18, in *Industrial Unrest in Great Britain,* Reprint of the Report of the Commission of Inquiry into Industrial Unrest (U.S. Department of Labor: Bureau of Labor Statistics No. 237), p. 119.
159 The British and Colonial Aeroplane Company Minute Book, No. 1, 23 October 1917.

the unity between workers in this particular case occurred within the walls of the company rather than without. The resolution of the long-running local company-union wrangle over the productivity bonus was possible because it could be contained within the bounds of the aircraft works.

In November, at a meeting on 'National Reconstruction' in St. Andrews Bristol, just a few weeks after the settlement of a four-day tramway strike in the city over union recognition,[160] Wethered delivered a sanctimonious speech about the emergence of class unity in the fight to defend democracy:

> The war was bringing to men and women a wider and fuller life than they had ever yet known. The war had given them an immensely broader outlook in the affairs of the world…trade unions, friendly societies, and co-operative societies had been called upon to assist the Government, a new sense of worth and dignity had come to large masses of ordinary people. In consequence of the war they had been driven to a new solidarity of national life. Class barriers had been broken down. The barrier of caste had been overthrown, and they had begun to understand as never before the meaning of democracy.[161]

Wethered used his speech to show that people needed to see that positive aspects had arisen out of the war effort and that the war aim of preserving British 'democracy' was worth fighting for. It was true that some welfare provision had been gained; on the other hand the State in wartime had assumed extensive powers to intervene in daily life. Many of these new measures were coercive rather than extensions of economic or social democratic rights. Even in terms of formal political rights, Wethered failed to mention that women, and indeed some working class men, did not have the right to vote in parliamentary elections, though the government had agreed to organise a 'representative conference' to examine electoral reform. In January 1917 its recommendations were submitted to the approval of the House of Commons but the passage of a Reform Bill had to wait over twelve months before it became an Act after Royal Assent.[162]

Wethered's 'democracy worth fighting' for address can be understood as camouflaging a desperate attempt to shore up the government's war objectives

---

160 See *Western Daily Press*, 22-26 October 10917; M. Richardson, *Trade Unionism and Industrial Conflict in Bristol: An historical study,* (Employment Studies Research Unit, University of the West of England), p. 25.
161 *Western Daily Press*, 21 November 1917, p. 5.
162 See J. D. Fair, 'The Political Aspects of Women's Suffrage during the First World War, *Albion: A Quarterly Journal Concerned with British Studies*, Vol. 8, No. 3 (Autumn, 1976), pp. 282-292.

at a time of militant trade union resistance in Britain, amid apprehension about the international threat posed by the October Russian Revolution. As Lloyd George recalled in his memoirs:

> The coming of the Russian Revolution lit up the skies with a lurid flash of hope for all who were dissatisfied with the existing order of society. It certainly encouraged all the habitual malcontents in the ranks of labour to foment discord and organise discontent.[163]

Despite the fears of the authorities of a national conflagration, industrial and political resistance was to manifest itself in more complex sporadic ways.

In Britain, during 1917, there had been a marked increase in the number of working days lost through strikes, despite the fact that they were illegal. In Bristol, the picture was mixed. The number of stoppages increased, but the number of working days lost fell compared with the previous year (see table 1 below).

Table 1: Number of stoppages* and aggregate number of working days lost, 1915—1918, in the Great Britain and Bristol.

Year	All industries and services in Great Britain		All industries and services in Bristol	
	No. of stoppages	Working days lost	No. of stoppages	Working days lost
1915	672	2,953,000	8	11,991
1916	532	2,446,000	8	39,846
1917	730	5,647,000	14	14,087
1918	1,165	5,879,000	21	48,997

*Excludes stoppages involving fewer than ten workers or lasting less than one day.
Source: *British Labour Statistics: Historical Abstract 1886-1968* (Department of Employment, HMSO 1971, p. 396; J. Love 'Some aspects of business and labour in Bristol during the First World War', (unpublished MA thesis, Bristol Polytechnic, 1986) p. 80.

When considering these figures account should be taken of the awards issued by the Arbitration Tribunals under the Munitions of War Acts to millions of workers; these aimed to placate workers and reduce the number of days lost per worker due to industrial disputes.[164] In Britain, many disputes were settled by arbitration and conciliation, and Bristol was no exception.

163 Lloyd George, *War Memoirs of David Lloyd George*, Vol. 11, pp. 1145-6.
164 Wolfe, *Labour Supply and Regulation*, p. 125.

However, because changes in the organisation of production had shifted the customary positions of male as well as female members of the workforce there were many possible sources of friction. Two examples from Bristol illustrate how prejudices towards women workers were still being expressed by male workers and by employers while at the same time women were developing a sense of entitlement to equality. In August 1917 the Society of Coppersmiths, Braziers and Metal Workers submitted an application. They wanted to transfer work done by women, employed at George Adams & Sons (Bristol), on oil coolers in the brass finishing shop, to men in the coppersmiths' shop. The union lost this appeal.[165] Then, in November 1917, the National Federation of Women Workers brought a claim against Parnall & Sons (Bristol) before the Special Arbitration Tribunal (For Women Employed on Munitions Work), for failing to pay the full wage entitlement to their female aircraft workers as prescribed by the Ministry of Munitions Orders issued on 28 January. In this case, brought against the employer, the union won and secured back pay, except in respect to women employed on woodcutting machines and as floor sweepers, whose wage rates were prescribed by different Orders.[166] State regulation combined with union action afforded women, who had been previously vulnerable at the workplace some room to manoeuver.

An unintended consequence of State intervention in industry was that economic discontent could actually focus on the State itself rather than on individual employers. On 13 December 1917, discontent over wage parity in Bristol came to a head when woodworkers engaged on aeroplane construction at the Brislington and Filton works of the B&CAC, and Parnall's, walked out to obtain the local enforcement of the national agreement that had secured a pay increase to compensate for the rise in cost of living. Their combined action brought the manufacture of aircraft in Bristol to a standstill with between 2,000 and 3,000 made idle by the strike.[167]

Earlier in the year metal workers and engineers in Bristol had secured a significant pay award to settle their grievances but the woodworkers had been left out. This response to the anomaly of conferring a 12½ per cent pay award to engineering and foundry trades and not to other crafts, or indeed less skilled workers, should not have come as a surprise to the Ministry. However, again consciousness proved to be convoluted and contradictory; for their complaint was not so much with their employers but with the Ministry of Munitions, the department responsible in government controlled establishments for pay

165 *Board of Trade Labour Gazette*, September 1917, p. 342.
166 *Board of Trade Labour Gazette*, December 1917, p. 469.
167 See the *Western Daily Press*, 14, 15 and 17 December 1917; The British and Colonial Aeroplane Company Minute Book, No. 1, 17 December 1917.

Women & Men working on the construction of Bristol F2B Fighter: Assembling Planes.

bargaining with union leaders.[168] Perhaps the Ministry may have been lulled into complacency because only a month before the workers had given an enthusiastic welcome to the King and Queen upon their visit to the B&CAC. A puzzled writer in *The Aeroplane* pondered over the enigma — why it was that this warm reception of the Royals by Bristol aircraft workers 'did not prevent the workers going on strike in December.'[169]

The authorities fearful for the war effort responded with remarkable alacrity to this dispute. Within a few days strikers received assurances that the claim would be settled in their favour if they returned to work immediately. At a meeting on Sunday 16 December they passed a resolution agreeing to do so 'on the understanding that the National agreement [would] come into force immediately.'[170] In February 1918, in an attempt to settle wage disputes, such as the one in Bristol, the issue of a Statutory Order established a legal minimum rate of pay and a standard working week for skilled woodworkers employed on aircraft construction.[171]

168 *Western Daily Press*, 14 December 1917, p. 5.
169 *The Aeroplane*, 9 January 1918, p. 159.
170 *Western Daily Press*, 17 December 1917, p. 5.
171 *The Official History of the Ministry of Munitions*, Vol. V, Pt. 2, Ch. V11, p. 89.

Women & Men working on the construction of Bristol F2B Fighter: Cabinet Shop.

Another interesting shift became evident too. By May, the numbers of women employed on metal work for aircraft were sufficiently high to force, somewhat belatedly, the issue of equal pay for equal work onto the agenda.[172] The Consolidated Women's Wages Order introduced on 8 May 1918 covered sheet metal and associated work that prior to the war had been customarily done by men. For this class of work, women on systems of payment by results were to receive the same piece-work rates as men (subject to deductions while workers were on probation or that skilled men customarily made ready the machines for women to operate). On types of work that had not been recognised before the war as men's work, women were to be paid according to the scale laid down in the women's work statutory Orders.[173]

Unrest, however, continued in 1918, in part due to the enforcement of the Ministry of Munitions ruling that 'skilled men must not be taken on to replace unskilled or semi-skilled men nor men to replace women'. This embargo

---

172 *The Official History of the Ministry of Munitions*, Vol. V, Pt. 2, Chapter V11, pp. 89-94: Drake, *Women in Trade Unions*, pp. 157-8.
173 *The Official History of the Ministry of Munitions*, Vol. V, Pt. 2, Chapter 11, p. 94.

Women Workers at Parnall's Coliseum Aircraft Works, 1918.

was introduced to prevent firms from hiring craftsmen, lured by higher wages and better conditions to replace workers, distinguished as unskilled or semi-skilled, conscripted to the Army. The turn around in status was a pragmatic adjustment to wartime conditions, which skilled workers' freedom of movement had made possible since the abolition of leaving certificates.[174] The 'Embargo Scheme', directed at skilled men, was suspected by the trade unions as being an underhand attempt to reintroduce industrial conscription.

As well as opposition to industrial conscription, resistance to new manpower proposals spiralled. Mass meetings called for an international peace conference and men gave pledges of defiance. The government's decision to cancel the certificates of exemption granted on occupational grounds, and to revise the Schedule of Protected Occupations, to clear the way for a further comb-out of munitions workers into the army, triggered a combative response from the shop stewards.[175] On 5 and 6 January, the National Administrative Council of the Shop Stewards and Workers' Committee met in Manchester and adopted a policy of active resistance to the further combing-out plans and called on the government to accept the Russian government's invitation 'to consider peace terms'. The Conference recommended 'that National action

174 *The Official History of the Ministry of Munitions*, Vol. V1, Pt. 2, Ch. 1V, p. 61.
175 *The Official History of the Ministry of Munitions*, Vol. V1, Pt. 2, Ch. 111, p. 39.

shall be taken to enforce these demands'. They instructed their delegates to consult their members as to 'what form this action shall take.'[176] Other unions adopted similar resolutions and a period of agitation for strike action to force the government to withdraw their manpower proposals followed. Convergence between militancy at the workplace and political opposition to the war was thus occurring. Lloyd George's government worst nightmare was turning into reality.[177]

The government's refusal to negotiate with the ASE in isolation, accusing the Engineering Union of wanting preferential treatment for its members, heightened tensions.[178] Concomitant with this rebuff, resolutions poured in from ASE District Committees and Union branches across the country opposing the government's manpower proposals along with calls for a negotiated peace. Bristol had eight branches with an aggregate membership of 2,500.[179] The resolution sent from Bristol No. 6 branch of the Union made explicitly left-wing political connections between their grievances at the workplace, peace and the redistribution of wealth:

> That the members of the Bristol 6th A.S.E. are resolved to resist the manpower proposals of the Government and demand that the Government shall at once accept the Russian Government's invitation to consider peace terms. They also demand the immediate conscription of wealth, and that adequate provision shall be made, as a right, for all victims of the war.[180]

The Shop Stewards National Conference had passed a similar resolution.[181]

On 14 January 1918, knowing it had the support from the majority of Labour representatives in Parliament, the government placed its proposals, as a Manpower Bill, for debate before the House of Commons. There followed a period of uncertainty as to the trade union response, especially the engineers.

On 27 January, the ASE London District Vigilance Committee called a meeting of its members to debate the Bill. Ten thousand responded, filling London's Albert Hall while four thousand gathered outside at an overflow meeting. James Hinton has contended that this combination of industrial unrest, hostility to the Manpower Bill and growing calls for a negotiated peace,

---

176 *Daily Herald,* 12 January 1918, p. 8; Hinton, *First Shop Stewards Movement*, p. 256.
177 K. Middlemas, *Politics in Industrial Society: The Experience of the British System since 1911* (André Deutsch, 1979), p. 115.
178 Hinton, *First Shop Stewards Movement*, p. 255.
179 *Western Daily Press*, 18 July 1917, p. 3.
180 *Daily Herald*, 2 February 1918, p. 6.
181 *Daily Herald*, 2 February 1918, p. 6.

particularly by the Women's Peace Crusaders, was positioning the leaders of the rank-and-file movement 'for revolutionary action against the war.'[182]

After a series of contributions, the Albert Hall meeting carried, with unanimous approval, the following demand:

> That the British Government enter into immediate negotiations with the other belligerent Powers for an armistice on all fronts, with a view to arranging a general peace on a basis of self-determination of all nations, no annexations, and no indemnities. Should such action demonstrate that German Imperialism is the only obstacle to peace, we express our determination to co-operate in the prosecution of the war until these objects are achieved. Failing such action on the part of the Government, we pledge ourselves to act with the organised workers of Britain in resisting man-power proposals of the Government. We further demand acceptable Labour representation of all countries at the proposed International Conference in order to ensure a people's peace.[183]

The careful wording made for a broad-based support. On the same day the Mersey District Engineering and Shipbuilding Trades Federation convened a meeting of 6,000 men to hear Sir Auckland Geddes explain the manpower proposals on behalf of the government. He failed to get his message across. Before Geddes had left the building, he witnessed the passing of a tougher resolution that pledged the Federation:

> to resist by all possible means any further call upon the remaining manpower of the nation, unless the Government immediately intimate their willingness to adopt the war aims as laid down by the Labour Party, permit and facilitate the holding of an International Workers' Conference, and agree without delay to conscript wealth.[184]

The following day Arthur MacManus and Willie Gallaher at a meeting in Glasgow proposed and seconded a resolution along similar lines ending with the words

---

182 Hinton, *First Shop Stewards Movement*, p. 256. Bristol was one of seventy-two branches of the Women's Peace Crusade. See J. Liddington, *The Long Road to Greenham: Feminism & Anti-Militarism in Britain since 1820* (Virago, 1989) for the origins of the Peace Crusade.
183 *Daily Herald*, 2 February 1918, p. 6.
184 *Daily Herald*, 2 February 1918, p. 6.

that the expressed opinion of the workers of Glasgow, is that from now on, and so far as this business is concerned, our attitude all the time and every time is to do nothing in support of carrying on the war, but to bring the war to a conclusion.[185]

Over the following week petitions for peace poured into the *Daily Herald* office.[186] The Manpower Bill became the centre of discussion at union branch meetings. Nevertheless, the subsequent failure to carry out the demands expressed at mass meetings of effective widespread strike action against the war seriously weakened the effectiveness of the main workers' organising body, the National Administrative Council of the Shop Stewards', and Workers' Committee Movement.[187] Never homogenous, divisions between craft and non-craft unions, disabled their ability to act as a collective force to oppose the Bill. Left-wing leaders in the shop stewards movement operated in a paradox for it was the skilled workers with privileges to defend who were the most intent on securing control over labour processes and often inclined to political radicalism. In 1918, apart from members of the ASE, particularly its younger element, most workers belonging to other unions rolled over and largely accepted the new scheme. Some turned against the ASE.[188]

With a few minor exceptions, the revision of the schedule of protected occupations raised the age when protection of men fit for enlistment began to twenty-three years. In the remaining occupations the age limit varied according to the importance of the work.[189] This further divided the unions. On 4 February, a lengthy discussion on the Bill took place at the Bristol Sheet Metal Workers' Union branch meeting, which culminated in approving 'that a strong deputation be appointed to wait upon the Ministry', to obtain the same age limit fixed for other sections of the Engineering Trades.[190] In short it was backing its union's position of supporting the manpower proposals as long as it secured the same conditions as the ASE. Two days later the Bill received Royal Assent.

The passing of the Bill did not stop a ballot of the ASE membership which resulted in the rejection of the government's new manpower directive by a

---

185 *Daily Herald*, 2 February 1918, p. 6; see also Hinton, *First Shop Stewards Movement*, p. 260.
186 *Daily Herald*, 9 February 1918, p. 3.
187 See Hinton, *First Shop Stewards Movement*, for a full account of the procrastination of the National Administrative Council of the Shop Stewards' and Workers' Committee Movement in enforcing the demands expressed at mass meetings.
188 *The Official History of the Ministry of Munitions*, Vol. VI, Pt. 2, Ch. 111, p. 45.
189 *The Official History of the Ministry of Munitions*, Vol. VI, Pt. 2, Ch. 111, p. 42.
190 Minutes of the General Meeting of the Bristol Operative Tin and Iron Plate, Sheet Metal Workers and Braziers, 4 February 1918.

majority of over 93,000. However, this fell short of the two-thirds majority of the total membership required to make the decision binding.[191] While the threat of a strike remained, the failure to act speedily according to the mandate of the membership meant the critical moment had been lost. The news of a German offensive, which began on 21 March, and the resurgence of patriotism that followed, finally extinguished any threat of a mass strike with the political ending of the conflict as an aim.[192]

Other causes of discontent nevertheless persisted; resistance to payment by results by aircraft woodworkers, the push for higher wages and the increasing discrepancy between the wages paid to women and men. Moreover, there was concern over health and safety at work. The Bristol Eye Dispensary had registered 3,000 new patients over the course of 1917 and reported that 'much of their work was for young women and men who were engaged at the aeroplane works and in munition factories.'[193]

During the summer of 1918, the Ministry of Munitions was receiving reports of restlessness among workers in Bristol, Manchester and London. In June, the chief investigation officer for Bristol and the South West reported:

> There has been and is continual agitation and unrest among the woodworkers and the claims and counter-claims of the carpenters and joiners, the shipwrights and boat builders to various classes of aircraft work, [that] keep the whole of those trades in a perpetual turmoil, and have a deleterious effect on output, and until these matters are settled by conference or arbitration I see no hope of better conditions or a lessening of the present inclination to strike on the slightest provocation.[194]

And in mid-July, the managing director of the B&CAC, Sir G. Stanley White and the company director secretary, H. White Smith travelled to London to meet with Sir Stephenson Kent, a member of the Council of the Ministry of Munitions, to discuss the 'Labour difficulties' the company was experiencing at its Filton Works.[195]

In Coventry, opposition to the Embargo Scheme led to an engineers' strike during July, involving men and women. Sympathy stoppages followed in Birmingham, South Staffordshire and East Worcestershire. The Bristol No. 5

---

191 *Western Daily Press*, 23 February 1918, p. 6.
192 Hinton, *First Shop Stewards Movement*, p. 266.
193 *Western Daily Press*, 12 February 1918, p. 3.
194 *The Official History of the Ministry of Munitions*, Vol. VI, Pt. 2, Ch. IV, p. 74.
195 The British and Colonial Aeroplane Company Minute Book, No. 1, 19 July 1918.

branch of the ASE also backed the Coventry workers who had gone against the advice of their national leaders and ceased work of their own accord. The Bristol branch forwarded a motion they had passed to the Ministry of Munitions saying:

> The members of Bristol 5th Branch of the Amalgamated Society of Engineers deplore the autocratic action by [the] Ministry in placing [an] embargo on skilled engineers and demand its immediate removal failing which they will support Coventry in their action.[196]

The 'moderate' John H. E. Flowers, the Bristol District secretary of the ASE and an assessor on Munitions Tribunals, said that his members had 'no objections to the rationing of skilled men as long as the employers and men are consulted in the rationing' and expressed his belief that 'unfortunately our only chance of getting justice is to strike.'[197] Even though this was a pale echo of the militant Glaswegian demand in 1915 for control over changes in the labour process, it reveals the pressure that must have been put upon Flowers.

There was a real chance that strikes against the Embargo Scheme would spread across the country. To head off this prospect the government did not resort to prosecuting engineers for illegal wartime striking but used a combination of carrot and stick devices. On the one hand the government offered to establish a Committee of Inquiry into the Embargo Scheme with union participation, while threating 'to withdraw from strikers protection from military service'.[198] The case for continuing the strike was not helped by negative responses from the public and the fact that semi-skilled and unskilled workers were not directly affected by the Scheme.

The *Western Daily Press* published one extreme view on the spectrum of public opinion, sent by a reader from Bishopston in Bristol, which painted the younger men working on munitions in a bad light:

> Let the Government take the thousands upon thousands of young fellows who are receiving (not earning) £4, £5, £6 and £7 a week, for being in a munition factory, and put them in the army, and we older men (I am 64) will give our services to do any work that our age will permit us to do…

196 *Western Daily Press*, 25 July 1918, p. 4.
197 *Western Daily Press*, 26 July 1918, p. 3.
198 *The Official History of the Ministry of Munitions*, Vol. V1, Pt. 2, Ch. 1V, p. 67.

The munition workers (etc.) will spend this on drinks in five minutes. Why does not the Government act in a firm manner, and conscript every man in the country up to the age of 65, put those young fellows into the army, and allot work to older men according to age?[199]

He was not alone. The following day three letters of support for the 'Bishopston' correspondent appeared.[200] It is impossible to ascertain how extensive such extreme attitudes were, though they certainly did not affect the strategically cautious approach taken by the government or the trade unions' responses.

The end of the Coventry strike and other stoppages came when on 29 July union officials accepted the government's offer of the establishment of a Committee of Enquiry into the Embargo Scheme with union participation. James Hinton describes how the call-up threat was the deciding factor. Indeed, on 29 July the Coventry strikers 'scuttled back to work in large numbers even before a mass meeting had accepted the settlement.'[201] On 31 July, the works' manager of the B&CAC was given the authority 'to engage certain classes of labour' following the government's withdrawal of the Embargo Scheme.[202]

Increasingly, planning for industrial reconstruction after the war was given priority and a number of conferences were held in Shipham, in Somerset, between Bristol employers and leading male trade unionists during 1918. The position of women in industry was one of the questions discussed: ominously the Bristol Association for Industrial Reconstruction decreed without any equivocation.

> the position of woman as industrial worker is and always must be of secondary importance to her position in the home. To provide the conditions which render a strong and healthy family life possible to all is the first interest of the State since the family is the foundation stone of the social system.[203]

The early unease expressed by traditionalists about women's commitment to the war effort was compounded by prejudices about women's sexual mores inside and outside the workplace. In February 1918, the Bristol and South West Counties Vigilance Association presented its annual report at a meeting of its membership, part of which voiced concern over the future trajectory of the moral

199 *Western Daily Press*, 3 April 1918, p. 2.
200 *Western Daily Press*, 4 April 1918, p. 2.
201 Hinton, *The First Shop Stewards' Movement*, p. 233; and for a full account of Coventry embargo strike see *ibid.*, p. 229—34.
202 The British and Colonial Aeroplane Company Minute Book, No. 1, 12 September 1918.
203 *Western Daily Press*, 10 July 1918, p. 3.

health of women factory workers during wartime. It concluded, pessimistically, with the remark 'it may be difficult and, perhaps, impossible to predict at the present time what its moral issue will be'. In the discussion following the report, Clifton College headmaster, Dr John Edward King, expressed extreme concern about the dangers faced in this new environment by 'young people of both sexes' calling 'upon all educational and religious institutions to do their utmost to train the coming generation in higher standards of purity and honour.' However, his admonition excluded what he called 'sex teaching.' On leaving school he wanted young people 'pledged to the new chivalry of personal purity and the suppression of the baser animal appetites.'[204]

His liberal concern about the moral development of working class women was expressed in relatively mild terms and was shared by many middle class men and women. It was accompanied however, by a powerful emotional backlash driven by a perception that women had acquired new freedoms in wartime. In June 1918, when the proprietor of *The Vigilante* (formerly the *Imperialist*), the right-wing Noel Pemberton Billing MP, was sued for libel by the well-known Canadian-born dancer Maud Allan for accusing her of lesbianism. Amid a blaze of publicity he was found not guilty. The verdict, as Lucy Bland observes, 'sent out a message equating sexual conformity with patriotic Englishness'.[205]

Angela Woollacott reported that to 'some observers, there was no doubt that World War 1 was a time of loosening of sexual mores in Britain.'[206] However, the assumption that this was a consequence of women joining men in factory work, as King from the remote Vigilance Association implied, was hardly realistic given the sexual vulnerability of working class women who had been employed previously as domestic servants, laundry and shop workers. Over the course of 1918, women workers, whose pay had marginally improved during the war, were to be caught up in a strong conservative reaction which had several manifestations. A rhetoric of both protection and moral outrage concealed a desire to return women to their pre-war position in society.

The Armistice

The gender backlash and patriotic reaction ran parallel with an opposing optimism in the possibility of social reconstruction. In late September, another weekend conference of the Bristol Association for Industrial Reconstruction got down to what they believed were the most important subjects regarding the

204 *Western Daily Press*, 15 February 1918, p. 3.
205 L. Bland, *Modern Women on Trial: Sexual Transgression in the Age of the Flapper* (Manchester University Press, 2013), p. 18.
206 Woollacott, *On Her Their Lives Depend*, p. 144.

rebuilding of industry once the war was over. The questions considered vital to industry included 'State control and State ownership, the functions of capital and labour, machinery for the better settlements of industrial disputes, national and local, [and] the new Education Act'.[207]

Just over one month later, Monday, 11 November 1918, the announcement that the armistice had been signed between the Allies and Germany was greeted with joy and relief. This news was instantly conveyed to the British & Colonial workforce at Filton and Brislington. Immediately, all overtime was stopped and a three-day holiday given to the whole of the workforce.[208] On 26 November, the Ministry of Munitions gave notice to the company of the termination of contracts of all Bristol Fighter spares. Moreover, a little later the company was notified that the contract to build Bristol Fighter planes had also been cancelled.[209] Further cancellations followed in the New Year. Though not well documented, significant numbers of men from the factory floor and most of the women transferred out of Bristol's aircraft factories in 1919 as jobs dried up.[210]

As a consequence by the middle of January 1919 there were over 2,500 women unemployed in Bristol. Immediately, in order to reduce the number of women claiming out-of-work benefit, pressure was put on these women to take up domestic service work.[211] This was applied energetically. Yet in April there were still 2,500 women on the unemployed books despite the fact that those women who rejected job offers had had their benefit payments withdrawn and were no longer considered as unemployed. It was reported in the local press that women 'factory workers, who have had domestic experience, are disinclined to accept domestic service again, and they are being struck off the register accordingly.…They have become accustomed to liberty, and now want the evenings free.'[212] Moreover, benefits were also withdrawn from those women who refused offers of employment 'out of town'. Patriotic service was swiftly forgotten.

Men too suffered with 8,000 unemployed in the city in May.[213] Some of these placed the blame for their predicament on what one correspondent described as 'muddling authorities who talk-talk-talk of reconstruction and make no effort

---

207 *Western Daily Press*, 2 October 1918.
208 The British and Colonial Aeroplane Company Minute Book, No. 1, 15 November 1918.
209 The British and Colonial Aeroplane Company Minute Book, No. 1, 12 December 1918 & 29 January 1919.
210 J. Roach the union representative of Amalgamated Society of Carpenters and Joiners was presented a leaving gift from his comrades on his transference from the Filton Works' propeller department, *Western Daily Press*, 28 January 1919, p. 4.
211 *Western Daily Press*, 18 January 1919, p. 4.
212 *Western Daily Press*, 24 April 1919, p. 4.
213 *Western Daily Press*, 14 May 1919, p. 6.

British & Colonial Aeroplane Company, Brislington Workers, 1918.

to reconstruct, who have taken men's jobs away for cheap female labour'.[214] By August unemployment for women in Bristol totalled around 1,700 and for men just over 7,000.[215] While these figures showed a slight improvement from those in the spring, for women in particular the prospects did not look promising due to the passing of the Restoration of the Pre-war Practices Act of August 1919 (forcible for one year) compelling employers to reinstate returning servicemen in their old jobs at the expense of substitute women and to a much lesser degree substitute men.[216] Pressure on women came from all sides. The media joined the campaign to cajole women back towards what was considered women's work in the pre-war period. Moreover, in May 1920, the local newspaper the *Western Daily Press* propounded the view

> that it should be a point of honour with girls whose fathers can afford to keep them at home, or who would not be inflicting any special hardships on themselves and their parents by being at home…they should remain at home for a little while.[217]

Recession, rising unemployment and government policy in the early 1920s meant that nationally many women were forced to take work as domestic servants. despite the efforts of women workers' organisations to keep equality on the agenda, women's labour rights were submerged.[218]

Bristol Aircraft Companies in 1920

In March 1920 the British & Colonial Aeroplane Company Ltd. went into voluntary liquidation following which it transferred all of its fixed assets to a new set-up, the Bristol Aeroplane Company, thereby avoiding the payment of the full imposition of the Excess Profits Duty (profits made during the war in excess of a pre-war standard of profits). Its operations turned towards supplying aircraft, such as the Bristol Pullman triplane, for newly opened commercial air-routes as well as continuing with government contracts.[219] A change of name, however, did not stop enquiries into the company's employment practices.

214 *Western Daily Press,* 1 May 1919, p. 6.
215 *Western Daily Press,* 28 August 1919, p. 5.
216 *The Official History of the Ministry of Munitions,* Vol. V1, Pt. 2, Ch. V, p. 92. For an examination of the importance of this Act, see G. R. Rubin, 'Law as a Bargaining Weapon: British Labour and the Restoration of pre-War Practices Act 1919', *The Historical Journal*, 32, 4 (1989), pp. 925-945.
217 *Western Daily Press,* 8 May 1920, p. 7.
218 Rowbotham, *Hidden from History*, pp. 116-119.
219 *Western Daily Press,* 3 September 1923, p. 7.

Women & Men working on the construction of Bristol F2B Fighter: Final Erecting.

Notably, given the Bristol Aeroplane Company's directors had stated, during the war, that they would retain women on aircraft work in peacetime, their policy of employing women was raised in the House of Commons in 1921. The Liberal MP, Atheistan Randall, asked the Secretary of State for Air whether there was any truth in the allegations that the company had

> discharged men workers and have reengaged women who were fully employed elsewhere, some women with husbands in employment; that such dismissed men are workless and have been compelled to seek help from the Lord Mayor of Bristol's Fund; and that as contractors to the Air Ministry this company is bound by the fair wage Clause'.[220]

He went on to call for the government to press the company to stop this practice. However, his call was rejected on the grounds that these women were not engaged on men's work. They were employed in the trimming department on machining and sewing of fabric as well as 'doping' and other light work, thus

---

220 *Hansard*, (20th Century House of Commons *Hansard* Sessional Papers) Fifth Series, Vol. 139. Col. 356-428, 28 April 1921.

the company was deemed to be conforming to ordinary standards of the Fair Wages Clause. This clause, which was included in all government contracts, required contractors to pay wage rates agreed between employers and workers' representatives and had originally been introduced as a result of campaigns against low pay.

The changed circumstances of peacetime led to contraction and amalgamation in the aircraft industry. In 1919 Cosmos Engineering purchased the aviation parts of Brazil, Straker and Company. However, in 1920 the company became insolvent, due to the lack of capital and the drying up of government contracts, and, under pressure from the government, was taken over by the Bristol Aeroplane Company. Parnall & Sons too lost government contracts after the armistice. In 1919, its managing director, George Parnall broke with the parent company, W & T Avery, and established a shop fitting and aircraft manufacturing business under his name at the Coliseum Works in Park Row.[221]

In January 1921, the *Western Daily Press* published an article on what it deemed as a 'vanished industry', the manufacture of aircraft in Britain: 'It is an unfortunate fact that much of the special skill, acquired by thousands of workers to meet the aircraft needs of the war period is now useless.' It cited the Filton aircraft works as an example where the building of motor bodies had, by 1921, largely replaced aircraft production.[222]

Summary and Conclusion

During the First World War workers were confronted by the incorporation of the trade union and labour leadership into the State machinery; the prohibition of strikes and lockouts; the introduction of compulsory arbitration; and a growth in State regulation over the wage rates of women and semi-skilled and unskilled men. The expansion of the armaments industry led to the large numbers of men and women being drawn away from their homes from disparate areas of the country to work in munitions centres. Women and semi-skilled and unskilled men either substituting for craftsmen or working on newly routinised and standardised production lines, such as aircraft assembly, gained greater recognition and status in relation to time-served skilled workers that precipitated a closing of the gap in wages between non-craft and craft workers.

Opposition to the effects of these changes erupted at differing times and in various centres of wartime production. It took the form of resistance to the erosion of customary boundaries between skilled and unskilled through

---

221 See J. Penny, *Bristol at Work*, pp. 178-185, for a full account.
222 *Western Daily Press*, 21 January 1921, p. 4.

dilution. It was directed at the introduction of women workers as well as unskilled and semi-skilled men. At the same time, skilled workers especially opposed conscription. These workplace rebellions gave rise to the shop stewards movement, workers' committees and the concomitant strengthening of workers' bargaining power. However, the impact of these movements from below was affected by conflicts of interest between differing groups within them and problems of coordination.

When the war ended, social discontent briefly focused on radical change, but was also accompanied by political and cultural pressure for a return to pre-war gender relations. While the manner in which these national developments occurred has been mapped out in many munitions centres, the specifics of what happened in Bristol have been neglected. This study shows how British & Colonial Aircraft Company expanded during the First World War, and documents the experience of the diversification of Parnall and Sons, and Brazil Straker, into the manufacture of aircraft, aero-engines and associated work.

The outbreak of war in August 1914 brought in its wake the demand for fighter aircraft in large numbers, on short lead times, that could only be met by significant investment in aviation infrastructure, and improvements in methods of production. A few Bristol companies took advantage of the opportunity to invest in new factories in the knowledge that lucrative government contracts to manufacture aircraft for use in the war would follow. As production expanded, and assisted by the Munitions Acts and State regulation, they made dramatic changes in work organisation particularly through the introduction of scientific management, removal of demarcation restrictions and the substitution of women for men.

While Bristol's aircraft industry reflects many of the broader national patterns, a close examination of labour and gender relations in Bristol's aircraft industry fills a void in our knowledge. Moreover, it illuminates how consciousness shifted over the course of the war. Bristol workers were initially moderate in contrast to the more highly political areas such as Glasgow. Nevertheless as a result of wartime measures, particularly the abolition of the trade card scheme, persistent labour unrest was to become a feature in Bristol's aircraft industry during the latter part of the war. Shop stewards emerged as important representatives of the rank-and-file securing leading positions on the B&CAC works committee. A collective consciousness began to develop. Despite the lack of individual testimonies, there is some evidence of the participation of men in the militant labour struggles becoming politically aware and later moving into Bristol's Labour politics.

This account also reveals the paradoxical position of women workers in wartime who on the one hand could enjoy living and supporting themselves independently from their parents and yet who also faced long hours of arduous and sometimes dangerous work, as well as having to deal with the hostility expressed sometimes by employers and by those men fearing dilution and the use of substitute women to lower wages. The interview with Elsie Hilliar offers an individual voice of a woman worker which is rarely heard. Her account of individual acts of sabotage suggests a mood of growing disillusionment despite patriotic calls to increase productivity. There had already emerged, as recent research has uncovered, a movement in Bristol against the war in which men and women participated.[223] It is evident that this was accompanied not only by an ethical stance on peace but also by left politics. A mix of war weariness and political hope was to trigger the call from leading branches of the Engineering and Metal workers' unions in Bristol to resist the Government's man power proposals and accept the offer to join Russia's peace talks with Germany in February 1918.

The armistice in November 1918, however, brought about a sudden and profound reversals in the trajectory of both labour and gender relations. As the requirement for armament production rapidly declined aircraft workers were laid off, unemployment rates rose considerably and the modest gains made by women through employment in Bristol's aircraft and munitions industry during the war were dramatically reversed. Although a period of intense social and industrial unrest occurred in Bristol after the end of hostilities in 1918, the collapse in the economy in late 1920 spawned a reactionary backlash.[224] Trade union membership declined substantially. Tarnished by its cooperation with the coalition government during the war, leaders of the local labour movement were unable to provide a resolute organised response to the injustice suffered by men and women workers and demobilised servicemen after the war. Interestingly, Samson Bryher, in his *Account of the Labour & Socialist Movement in Bristol*, observed the psychological effects: 'Of those who went and were fortunate enough to return, many had developed jagged nerves.' They were left workless and disillusioned.[225]

Nevertheless, in the post-war period many Bristol workers, unable to secure the economic gains of the war by industrial militancy, turned to the political wing of the Bristol labour movement to promote working class interests.[226]

223 See Thomas, *Slaughter No Remedy: The Life and Times of Walter Ayles* and Hannam, *Bristol Independent Labour Party: Men, Women and the Opposition to War.*
224 K. Kelly and M. Richardson, 'The Shaping of the Bristol Labour Movement, 1885-1985' in M. Dresser and P. Ollerenshaw (eds), *The Making of Modern Bristol* (Redcliffe Press, 1996) pp. 219-220.
225 S. Bryher (Samuel Bale), *An Account of the Labour & Socialist Movement in Bristol* Part III (Bristol Labour Weekly, 1929), p. 7.
226 Kelly and Richardson, 'The Shaping of the Bristol Labour Movement, 1885-1985', p. 221.

In the 1920s, Labour representation on Bristol City Council increased substantially largely at the expense of the Liberals. In December 1923, Walter Ayles, the former reviled conscientious objector, became one of Bristol's first Labour MPs.[227] His constituency, in which some munitions factories and aircraft works had operated during the war, was confronted with the formidable problem of over capacity for peacetime armament needs. Employment in the British aircraft industry had collapsed from a wartime high of over 250,000 to less than 12,000 in 1924.[228] Employment on aircraft production in the Bristol Aeroplane Company also fell substantially. However, the company's turn to producing aero-engines, after its take-over of Cosmos Engineering in 1920, gave it new life. Employment on Bristol Aeroplane's engine division increased from a staff of forty-two in 1920 to around 3,000 in 1935. The expansion of the Bristol engine division was greatly assisted by subsidies and orders from the State, and the advent of the British government's rearmament programme.[229] The revival of the aircraft industry brought with it the emergence of a new wave of industrial unrest and the building of a network of rank-and-file militants.[230]

---

227 S. Jordan, K. Ramsey and M. Woollard, (Series Editor P. Wardley), *Abstract of Bristol Historical Statistics Part 3: Political Representation and Bristol's Elections 1700-1997* (University of the West of England, 1997). Ernest Bevin, who was forthright in condemning militarism before the outbreak of war, but cooperated in the war by encouraging dockers to keep goods and munitions moving swiftly through British ports, unsuccessfully contested the Bristol Central seat for Labour in the 1918 General Election. James Kaylor, Labour's General Election candidate for Bristol North in 1918, was also unsuccessful. In May 1920, a leading anti-war activist, and Independent Labour Party member, Mabel Carole Tothill, became the first woman to be elected on to Bristol City Council, although she lost her seat eighteen months later (Hannam, *Bristol Independent Labour Party*, p. 50).
228 D. Edgerton, *England and the Aeroplane; An Essay on a Militant and Technological Nation* (Macmillan, 1991), p. 26.
229 W. S. Hornby, *History of the Second World War: Factories and Plant* (HMSO, 1958), p. 251; G. Stone, 'Rearmament, War and the Performance of the Bristol Aeroplane Company, 1935-45' in C. Harvey and J. Press (eds.) *Studies in the Business History of Bristol* (Bristol Academic Press, 1988), p. 189; R. Schlaifer and S. D. Heron, *Development of Aircraft Engines and Fuels: Two studies of relations between government and business* (Boston: Division of Research, Graduate School of Business Administration, Harvard University, 1950), pp. 138 & 153 (appendix to Chapter V1). In this appendix it is revealed that between 1920 and 1930 'on average probably nine-tenths or more of Bristol [Engine] development was paid by the government, and at least two-thirds (usually more) of all sales were to the government'.
230 A. Danford, M. Richardson, P. Stewart, S. Tailby and M. Upchurch, 'The Legacy of Trade Union Power' in Danford, Richardson, Stewart, Tailby and Upchurch, *Partnership and the High Performance Workplace: Work and Employment Relations in the Aerospace Industry* (Palgrave Macmillan, 2005), p. 22-23.

Appendix

Card transferring Joseph Entwistle to Army Reserve, 6 July 1916.

Bristol Citizens' Recruiting Committee card stating that that Joseph Entwistle was medically unfit, 26 November 1915.

Medical classification certificate of Joseph Entwistle issued by the Medical Recruiting Board, Colston Hall, Bristol, 31 May 1916.

Card transferring Joseph Entwistle to the Army Reserve, 6 June 1916.

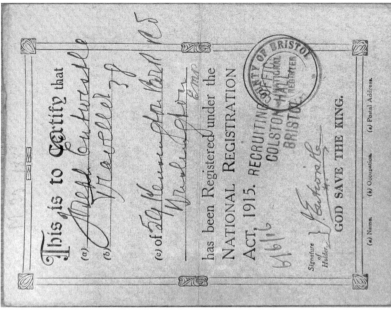

Registered under the National Registration Act, 1915, 6 June 1916.

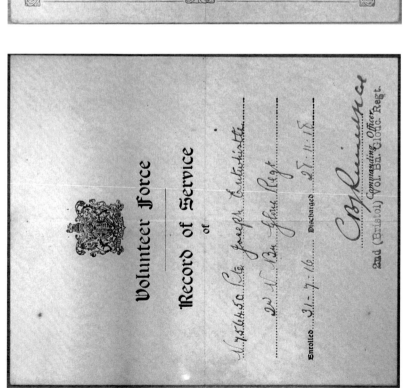

Volunteer Force Record of Service, 31 July 1916 to 28 November 1918.

[811] W13090/4672 750m. 11/15 G & S Forms/B. 2512/1.   Army Form B. 2512A.

SHORT SERVICE.

(For the Duration of the War with the Colours and in the Army Reserve.)

NOTICE to be given to a MAN at the time of his offering to join the Army.

Date _____ 191_

You (name) _____ J. Entwistle _____

are required to attend forthwith, or

at _____ o'clock _____ on the _____ day of _____

at _____ [here name some place]

for the purpose of appearing before a Justice to be attested for His Majesty's Army, in which you have expressed your willingness to serve.

The General Conditions of the Contract of Enlistment that you are about to enter into with the Crown are as follows:—

1. You will engage to serve His Majesty as a Soldier in the Regular Forces for the duration of the war, at the end of which you will be discharged with all convenient speed. You will be required to serve for one day with the Colours and the remainder of the period in the Army Reserve in accordance with the provisions of the Royal Warrant dated 20th October, 1915, until such time as you may be called up by order of the Army Council. If employed with Hospitals, Depôts of Mounted Units, or as a Smith, etc., you may be retained after the termination of hostilities until your services can be spared, but in no case shall this retention exceed six months.

2. Your term of Service will be reckoned thus:—
   1. The Service shall begin to reckon from the date of Attestation, but
   2. If guilty of any of the following offences:—
      (a) Desertion from His Majesty's Service, or
      (b) Fraudulent Enlistment,
   the whole period of Service prior thereto shall be forfeited.

3. You will be enlisted for General Service, and appointed to a Corps when called up for Service.

4. If serving beyond the seas at the time you are entitled to your discharge, you will be sent to the United Kingdom free of expense.

5. When attested by the Justice you will be liable to all the provisions of the Army Act for the time being in force.

6. You will be required by the Justice to answer the questions printed on the back hereof, and you are warned that if you make at the time of your attestation any false answer to him you will thereby render yourself liable to punishment.

Signature of the Non-Commissioned }
   Officer serving the Notice   } _____

* The Recruit is to have the option of being attested either forthwith, or at a future time. If he wishes to be attested forthwith, the words "or at ____ o'clock on the ____ day of _____" (in line 3) will be erased. If he does not wish to be attested forthwith, the hour (with the letters a.m. or p.m. as the case may be), the day, and the month will be inserted, and the words "forthwith or" (in line 2) will be omitted.

[TURN OVER.

Short Service Notice with Attestation stamp, 6 June 1916.

Army Form 2512A.

SHORT SERVICE.
(For the Duration of the War with the Colours and in the Army Reserve)

NOTICE to be given to a Man at the time of his offering to join the Army

Date...June 6th 1916......

You (name) J. Entwistle

are required to attend forthwith, or at Recruiting Office, Colston Hall, for the purpose of appearing before a Justice to be attested for His Majesty's Army, in which you have expressed your willingness to serve.

The General Conditions of the Contract of Enlistment that you are about to enter into the Crown are as follows: —

1. You will engage to serve His Majesty as a Soldier in the Regular Forces for the duration of the war, at the end of which you will be discharged with all convenient speed. You will be required to serve one day with the Colours and the remainder of the period in the Army Reserve in accordance with the provisions of the Royal Warrant dated 20th October 1915, until such time as you may be called up by order of the Army Council. If employed with Hospitals, Depôts of Mounted Units, or as a Clerk, etc., you may be retained after the termination of hostilities until your service can be spared, but in no case shall this retention exceed six months.
2. Your term of Service will be reckoned thus:
   1. The Service shall begin to reckon from the date of Attestation but
   2. If guilty of any of the following offences: —
      a. Desertion from His Majesty's Service, or
      b. Fraudulent Enlistment,
      the whole period of Service prior thereto shall be forfeited.
   3. You will be enlisted for General Service, and appointed to a Corps when called up for Service.
   4. If serving beyond the seas at the time you are entitled to your discharge, you will be sent to the United Kingdom free of expense.
   5. When attested by the Justice you will be liable to all the provisions of the Army Act for the time being in the force.
   6. You will be required by the Justice to answer the questions printed on the back hereof, and you are warned that if you make at the time of your attestation any false answer to him you will render yourself liable to punishment.

Signature of the Non-Commissioned Officer serving the Notice }

Any further communication should be
addressed to-

  THE SECRETARY,

 And the following letter
and number should be quoted:-
  M.S.14231.
Telegraphic Address: Munisupply, London.
Telephone Number: Victoria,8660-5
    Treasury,221-2

Labour Enlistment
Complaints Section.

MINISTRY OF MUNITIONS OF WAR,
6, Whitehall Gardens,
LONDON. S.W.

Gentlemen,

  I am directed to refer to this Department's letter of 26/2 relative to the retention of Jos. Entwistle in civil employment, and to state that the War Office have informed this Department that the Recruiting Officer concerned has been instructed not to call him up without further instructions from the War Office.

  It may, however, be necessary to call up this man again with a view to his formal transfer to Class W. Army Reserve, after enrolment, and his return to civil employment.

      I am, Gentlemen,
       Your obedient Servant,
        F.R. Lovett.

L.E.C. 13.

  Messrs. Pritchett & Gold.

Letter from Ministry of Munitions Labour Enlistment Complaints Section confirming that Joseph Entwistle is not to be called up pending further instructions.

Protected Occupation Certificate, 11 May 1917, Bristol Archives.

Certification of Exemption, Bristol Archives, 1 March 1917.

Bibliography

Archives

Bristol Archives, Joseph Entwhistle papers.
Bristol Central Reference Library.
Bristol Museums Galleries Archives
British Newspaper Archive, (FindmyPast Newspaper Archive Limited).
Imperial War Museum Sound Archive, Elsie Hilliar (interview by Chris Howell, 1982), Cat. No. 6682, Reel 1.
Maclean, John, Archives, (Marxists.org).
Richardson, Mike, Papers, (The Bristol Operative Tin and Iron Plate, Sheet Metal Workers and Braziers).
Royal Aeronautical Society Archives, RAES/BCAC/DIRE/1 (The British and Colonial Aeroplane Company Minute Book, No 1).
Royal Air Force Museum, rafmuseum.org.uk
University of Liverpool, Glasier, Bruce, Papers.

Parliamentary Papers, Government Publications, Official Reports and Registers

Board of Trade Labour Gazette
Census of England and Wales 1911, courtesy of the National Archives/Findmypast Limited 2015.
Hansard, (20th Century House of Commons Hansard Sessional Papers) Fifth Series, Vol. 139. Col. 356-428, 28 April 1921.
The Official History of the Ministry of Munitions Volumes 1-X11 (HMSO, 1920-24).
U.S. Department of Labor: Bureau of Labor Statistics No. 237, *Commission of Inquiry into Industrial Unrest*, No. 6 Division. Report of the Commissioners for the Southwest area, 1917/18, in *Industrial Unrest in Great Britain*, Reprint of the Report of the Commission of Inquiry into Industrial Unrest.

Newspapers and Periodicals

Aeroplane, The
Birmingham Mail
Bristol and the War
Bristol Times & Mirror
Coventry Evening Telegraph
Daily Herald
Flight
Hendon & Finchley Times
Manchester Evening News
The Times
Vanguard
West Bristol Labour Weekly, The
Western Daily Press

Web Sites

Gracesguide.co.uk
rafmuseum.org.uk
www.b-i-a-s.org.uk
aerosociety.com
lboro-history-heritage.org.uk

Unpublished theses

Batchelor, L. A., 'A Great munitions centre: Coventry's armaments and munitions industry 1914-1918.' (Unpublished MScR Thesis. Coventry: Coventry University, 2008).

Whitson K., 'Scientific Management Practice in Britain, A History' (unpublished PhD, University of Warwick, 1995).

Books, Articles and Pamphlets

Adams, R .J. Q., 'Delivering the Goods: Reappraising the Ministry of Munitions 1915-1916', *Albion: A Quarterly Journal Concerned with British Studies*, Vol. 7, No. 3, Autumn, 1975.

Alford, B. W., *Britain in the World Economy Since 1880* (Longman, 1996).

Aris, R., Trade Unions and the Management of Industrial Conflict (Macmillan Press, 1998).

Backwith, D. Ball, R. Hunt S. E. and Richardson M. (eds.) *Strikers, Hobblers, Conchies & Reds: A Radical History of Bristol 1880-1939* (London: Breviary Stuff Publications, 2014).

Barnes, C. H., 'Bristol and the Aircraft Industry', *Bristol Industrial Archaeological Society Journal* 34, 1972.

Bell, T., *Pioneering Days* (Lawrence & Wishart, 1941).

Bland, L., *Modern Women on Trial: Sexual Transgression in the Age of the Flapper* (Manchester University Press, 2013).

Bourke, J., 'Housewifery in Working –Class England 1860-1914' in *Past and Present*, 1994, 143 (1).

Brake, T., *Men of Good Character: A History of the Sheet Metal Workers, Coppersmiths, Heating and Domestic Engineers* (Lawrence and Wishart, 1985).

Braverman, H., *Labor and Monopoly Capital* (Monthly Review Press, 1974).

Braybon, G., *Women Workers in the First World War: The British Experience* (Routledge, 2013, first published 1981).

Bruce, J. M., 'The Bristol Scout: Historic Military Aircraft No. 18, Part 1', in *Flight*, 26 September 1958.

Bryher S. (Samuel Bale), *An Account of the Labour & Socialist Movement in Bristol* Part III (Bristol Labour Weekly, 1929).

Budd, L. C. S., 'Selling the Early Air Age: Aviation Advertisements and the Promotion of Civil Flying in Britain, 1911-1914', *Journal of Transport History*, 32. 2 Dec 2011.

Cadogan, M., *Women with Wings: Female Flyers in Fact and Fiction* (Macmillan, 1992)

Clegg, H. A., *A History of British Trade Unions since 1889: Volume 11 1911-1933* (Oxford University Press, 1985).

Creighton, C. 'The Rise of the Male Breadwinner Family: A Reappraisal' in *Comparative Studies in Society and History*, Vol. 38, No. 2, April 1996.

Currie, R., *Industrial Politics* (Oxford: Clarendon Press, 1979).

Danford, A., Richardson, M., Stewart, P., Tailby S., and Upchurch, M., 'The Legacy of Trade Union Power' in Danford, A., Richardson, M., Stewart, P., Tailby, S., and Upchurch, M., *Partnership and the High Performance Workplace: Work and Employment Relations in the Aerospace Industry* (Palgrave Macmillan, 2005).

Drake, B., *Women in the Engineering Trades* (Fabian Society & George Allen & Unwin, 1917).

Drake, B., *Women in Trade Unions* (Virago, 1984, first published in 1920 by the Labour Research Department).

Dresser M., and Ollerenshaw P., (eds), *The Making of Modern Bristol* (Redcliffe Press, 1996).
Driver, H., *The Birth of Military Aviation: Britain, 1903-1914* (Boydell & Brewer, 1997).
Edgerton, D., *England and the Aeroplane; An Essay on a Militant and Technological Nation* (Macmillan, 1991).
Fair, J. D., 'The Political Aspects of Women's Suffrage during the First World War', *Albion: A Quarterly Journal Concerned with British Studies*, Vol. 8, No. 3 (Autumn, 1976).
Freeston, C. L. 'The British Aircraft Industry' in *Aeronautical Engineering: supplement to "The Aeroplane"* 2 November 1917.
Gleason, A., *What The Workers Want: A Study of British Labour* (New York: Harcourt, Brace and Howe, 1920).
Hamilton, A. 'Dope Poisoning in the Manufacture of Airplane Wings', *Monthly Review of the U.S, Bureau of Labor Statistics*, Vol. 5, No. 4, October 1917.
Hannam, J., *Bristol Independent Labour Party: Men, Women and the Opposition to War* (Bristol Radical History Group, Pamphlet 31, 2014).
Hardach, G., *The First World War 1914-1918* (Pelican, 1987, first published in English by Allen Lane, 1977).
Hiley, N., ' Counter-espionage and security in Great Britain during the First World War', *English Historical Review*, 1986.
Hinton, J., 'Introduction' in Reprints in Labour History No. 1 to J. T. Murphy *The Workers' Committee: An Outline of its Principles and Structure* (first published by the Sheffield Workers' Committee 1917, Pluto Press, 1972).
Hinton, J., *The First Shop Stewards' Movement* (London: George Allen & Unwin, 1973).
Hornby, W. S., *History of the Second World War: Factories and Plant* (HMSO, 1958).
Howell, C., *No Thankful Village: The Impact of the Great War on a Group of Somerset Villages- a microcosm* (Bath: Fickle Hill, 2002).
J. Love 'Some aspects of business and labour in Bristol during the First World War' (unpublished MA thesis, Bristol Polytechnic, 1986).
James, D. J., *The Bristol Aeroplane Company* (Tempus Publishing, 2001).
Jordan, S. Ramsey K. and Woollard, M., (Series Editor P. Wardley), *Abstract of Bristol Historical Statistics Part 3: Political Representation and Bristol's Elections 1700-1997* (University of the West of England, 1997).
Kelly K., and Richardson, M., 'The Shaping of the Bristol Labour Movement, 1885-1985' in M. Dresser and P. Ollerenshaw (eds), *The Making of Modern Bristol*.
Lewenhak, S., *Women and Trade Unions* (Ernest Benn, 1977).
Liddington, J., *The Long Road to Greenham: Feminism & Anti-Militarism in Britain since 1820* (Virago, 1989).
Lloyd George, D., *War Memoirs of David Lloyd George*, Vol. 1 (Odhams Press, 1938).
Lloyd George, D., *War Memoirs of David Lloyd George*, Vol. 11 (Odhams Press, 1938).
Maclean, J., 'The Clyde Unrest', *Vanguard*, November 1915.

Mant, A. K., M.D., 'Acute Tetrachlorethane Poisoning' in *The British Medical Journal*, Vol.1, 21 March 1953.
Middlemas, K., *Politics in Industrial Society: The Experience of the British System since 1911* (André Deutsch, 1979).
Millman, B., *Managing Domestic Dissent in First World War Britain* (Frank Cass, 2000).
More, C., *Skill and the English Working Class, 1870-1914* (Croom Helm, 1980).
Murphy, J. D., 'Aircraft Production' in S. C. Tucker (ed.) *World War 1 Encyclopedia* (California: ABC-CLIO, 2005).
Murphy, J. T., 'The Shop Stewards and Workers' Committee Movement' in A. Gleason, *What The Workers Want*.
Murphy, J. T., *Preparing for Power* (Pluto Press, 1972, first published by Jonathan Cape, 1934).
Penny, J., *Bristol at Work* (Derby: Breedon Books, 2005).
Richardson, M. 'Bristol and the Labour Unrest of 1910-14' in D. Backwith, R. Ball, S. E. Hunt and M. Richardson (eds.) *Strikers, Hobblers, Conchies & Reds*.
Richardson, M., *Bliss Tweed Mill Strike 1913-14: Causes, Conduct and Consequences* (Bristol Radical History Group, Pamphlet 26, 2013).
Richardson, M., *Trade Unionism and Industrial Conflict in Bristol: An historical study*, (Employment Studies Research Unit, University of the West of England).
Rowbotham, S., *Friends of Alice Wheeldon: The Anti-War Activist Accused of Plotting to Kill Lloyd George* (Pluto, second edition, 2015, first published in 1986).
Rowbotham, S., *Hidden from History* (Pluto, third edition, 1977).
Rubin, G. R., 'Law as a Bargaining Weapon: British Labour and the Restoration of pre-War Practices Act 1919', *The Historical Journal*, 32, 4 (1989).
Schlaifer, R. and Heron, S. D., *Development of Aircraft Engines and Fuels: Two studies of relations between government and business* (Boston: Division of Research, Graduate School of Business Administration, Harvard University, 1950).
Stone, G., 'Rearmament, War and the Performance of the Bristol Aeroplane Company, 1935-45' in C. Harvey and J. Press (eds.) *Studies in the Business History of Bristol* (Bristol Academic Press, 1988).
Stephens, M., Ernest Bevin – *Unskilled Labourer and World Statesman 1881-1951* (Transport & General Workers Union, 1981).
Thomas C., *Slaughter No Remedy: The Life and Times of Walter Ayles, Bristol Conscientious Objector* (Bristol Radical History Group, Pamphlet 36, 2016).
Treadwell, C., *British & Allied Aircraft Manufacturers of the First World War* (Amberley, 2011).
Whitfield, R., 'Trade Unionism in Bristol 1910-1926' in I. Bild (ed.), *Bristol's Other History* (Bristol Broadsides, 1983).
Wixey, K., *Images of England: Parnall's Aircraft* (Tempus Publishing, 1998).
Wolfe, H., *Labour Supply and Regulation* (Oxford: Clarendon Press, 1923).
Woollacott, A., *On Her Their Lives Depend: Munitions Workers in the Great War* (University of California Press, 1994).
Wrigley, C., *Lloyd George* (Blackwell, 1992).

Name Index

Addison, Christopher MP, 33
Allan, Maud, 48
Ashmore, J. D., 24
Asquith, Herbert Henry, (Prime Minister 1908-16), 27
Ayles, Walter, 21, 55, 56
Barnes, George, 32
Bevin, Ernest, 34, 56
Biggs, L., 24
Bland, Lucie, 48
Brancker, Colonel, 10
Bryher, Samson, 55
Burgess, S., 31
Cassidy, N., 31
Clarke, Albert Edward, 10
Currie, Robert, 11
Dingley, T. W., 31
Entwistle (Entwhistle), Joseph, 28, 57-63
Flowers, John, 46
Gallaher, Willie, 43,
Geddes, Auckland, Sir, 43
Grey, Grey Charles, 18
Haldane, Richard (Viscount), 1
Headley, Lord, 19
Henderson, Arthur MP, 17-19, 32
Henderson, David, 9
Hill, W., 31
Hilliar, Elsie, 25-26, 28-29, 55
Hinton, James, 42, 47
Hodge, John, 32
Jackson, W., 24
Kaylor, James, 14, 17, 19, 24
Kealey, P. H., 31
Kent, Stephenson, Sir, 45
King, J. E. Dr., 48
Lewis, T. C., 34

Lloyd George, David, MP (Prime Minister Dec. 1916), 7-10, 12, 14, 17-19, 21, 27, 30, 32-33, 37, 42
Lord Derby, 16
Macarthur, Mary, 13-14
Maclean, John, 12
MacManus, Arthur, 19, 31, 43
Muir, John, 17
Parnall, George, 53
Peet, G., 31
Pemberton Billing, Noel, MP, 48
Piper, G., 24
Pole, Albert, 20
Princess Louise, H. R. H., 33
Randall, Atheistan, 52
Rowntree, Arnold MP, 34
Scottow, T. J., 34
Seddon, James Andrew, 13
Shiner, J. D., 24
Simon, John, Sir, 22-23
Smith, Allan, 14
Smith, H. White, 45
Swaish, John, 24
Taylor, L. T., 34
Tothill, Mabel Carole, 56
Wakeham, A., 24
Walsh, P., 24
Watson, W., 31
West, Glyn, 14
Wethered, Ernest Handel Cossham, 34, 36
White, G. Stanley, 10, 22-23, 45,
White, George, Sir, 1-2, 5, 7-8, 11
Wolfe, Humbert, 6, 9, 11
Woollacott, Angela, 13, 48

General Index

ACTs & Bills
   1914 Defence of the Realm Act, 4, 23
   1915 Munitions of War Act, 12, 14, 29, 37
   1915 Munitions of War Bill, 11
   1916 (Jan.) Munitions of War (Amendment) Act, 19, 20
   1916 Military Service Act, 22, 29
   1916 Military Service Bill, 20, 21
   1917 Munitions of War Act, 32
   1917 Munitions of War Bill, 32
   1918 Manpower Act, 44
   1918 Manpower Bill, 42
Admiralty, 2-3, 5, 15-16
Amalgamated Society of Carpenters and Woodworkers, 6, 49
Amalgamated Society of Engineers (ASE), 6, 13-17, 19, 21, 23-24, 30-32, 42, 44, 46
Amalgamated Society of Ironfounders, 24
Anti-conscription conferences and anti-war demonstrations, 18, 22-23, 56
Arbitration Tribunals, 37
ASE London District Vigilance Committee, 42
B.E.10 biplane, 3
B.E.2.Ds, 8, 10
B.E.2c reconnaissance aircraft, 3, 5, 6
B&CAC Filton Works Committee, 34-35, 54
B&CAC, see British & Colonial Aeroplane Company
Belgium, 18
Box-Kite biplane, vi, 2
Brazil Straker & Company Ltd., 1, 15-16, 53-54

Brislington Aircraft Works, 5, 7-8, 10, 23, 27, 38, 49, 50
Bristol Aeroplane Company, see also B&CAC,
Bristol Association for Industrial Reconstruction, 47-48
Bristol branch of the Committee on Women's War Employment,
Bristol Bullet, and see B.E.2c, 5
Bristol City Council, 56
Bristol City Museum and Art Gallery, 1,
Bristol Corporation, 30
Bristol Labour Exchange, 26
Bristol Labour Representation Committee, 21, 56
Bristol Munitions Court, 34
Bristol No. 5 branch of the ASE, 45-46
Bristol No. 6 branch of the ASE, 42
Bristol Operative Tin and Iron Plate, Sheet Metal Workers and Braziers, 7, 20, 44
Bristol Recruiting Office, 28
Bristol Scout, and see B.E.2c
Bristol Tractor biplanes, 3, 5
Bristol Trades Council, 21, 34
Bristol Tramways Company, 1, 8
Bristol Volunteer Force, Gloucester Regiment, 28
British & Colonial Aeroplane Company (B&CAC), 1-3, 5-10, 22-23, 25-29, 30, 34, 35, 38-39, 45, 47, 49, 50, 51, 54
Central Munitions Labour Supply Committee, 14,
Circular L.2, 15, 20
Circular L.3, 15
Clifton College, 48,
Clyde Workers' Committee, 12, 16-17

Clydeside, 16, 18, 19
Committee of Enquiry into the Embargo Scheme, 47
Compulsion, 11, 22
Conscription, 12, 18-24, 29-31, 41-42, 54
Consolidated Women's Wages Order (1918), 40
Controlled establishment, 10-12, 15, 19, 38
Cosmos Engineering, 53, 56
Coupon Election, 34
Coventry, 25-26, 31, 45-47
Dagenham, 28
Deputy director of Military Aeronautics, 8
Derby Scheme, 16, 19
Dilutee, 25
Dilution of labour, 6, 16
Dilution, 4, 6-7, 10, 14-21, 24, 27, 30-33, 54-55
  Substitution, 12, 15, 22, 31, 54
Director of Air Organisation, 10
Director-General of Military Aeronautics, 3
Dock, Wharf, Riverside and General Workers' Union, 15, 24
Doping, 26-28, 52
Embargo Scheme, 41, 45, 46-47
Engineering Employers' Federation, 14
Equal pay, 13-14, 40
Exemption, 24, 27-28, 41, 63
Fair Wages Clause, 53
Farnborough, 3, 11, 26
Filton Aircraft Works, 7, 10, 23, 25-26, 29, 38, 45, 49, 53
Filton Tramway Depot, 2
France, 5
G. & J. Weir Engineering Co., 6
Garrison Service, 28
George Adams & Sons (Bristol), 38
Germany, 3, 5, 49, 55
German Agents, 29

German Imperialism, 43
German Offensive, 45
German Plots, 18
German Troops, 18
Glasgow, 14, 16-18, 43-44, 54
Government-owned factories, 14
Hydroaeroplane, 2
Imperial War Museum, 25
Independent Labour Party, 14, 18, 21, 23, 34, 35, 56
Instructional Aircraft Factory, Gloucester Road, north Bristol, 30
King's Squad Scheme, 9
Labour Enlistments Complaints Committee, 28
Labour Party, 17, 34, 43
Lark Hill, Salisbury Plain, 2
Leaving Certificate, 12, 13, 32, 41
Liverpool, 31
Local Labour Advisory Committees, 24
London, 20, 31, 42, 45
Manchester, 31, 41, 45
Manpower Proposals, 41-44
Marne, Battle of, 5
Mersey District Engineering and Shipbuilding Trades Federation, 43
Minister of Munitions, 7, 10, 19, 33
Ministry of Munitions, 6, 11-12, 14, 15, 18, 20, 23-25, 27, 29, 31, 38, 40, 45-46, 49, 62
Ministry of Munitions Labour Intelligence Division, 29
Ministry of Reconstruction, 35
Ministry of War, 10
Munitions of War, 10, 11-12, 14, 19, 20, 24, 29-30, 32, 37
Munitions Tribunals, 27, 46
National Administrative Council of the Shop Stewards and Workers' Committee, 41, 44

National Advisory Committee on War Output to the Ministry of Munitions, 24
National Amalgamated Sheet Metal Workers, 6, 7, 11, 20, 44
National Federation of Women Workers, 4, 13-15, 19, 24, 27, 38
National Union of Railwaymen, 14
National Union of Shop Assistants, 13
North Wales Independent Labour Party Federation, 34
North-East Coast Armaments Committee, 9
Parnall and Sons, 1, 15-16, 20, 27, 29, 38, 41, 53, 54
Patriotism, 7, 18, 45
Pritchett & Gold Electric Power Storage Company, 27-28
Public Opinion, 19, 46
Radstock, 25
Rank-and-file, 3, 11, 19, 32, 43, 54, 56
Rolls-Royce, engines, 16
Royal Aircraft Factory, 3, 5, 6, 26
Royal Flying Corps, 6,
Russian Revolution (October 1917), 37
Sabotage, 18, 29, 55
Schedule of Protected Occupations, 41, 44
Scotland, 16, 20
Secretary of State for War, 1
Sheffield, 16, 31-32
Shop Stewards, 5, 14, 16-19, 30-32, 34-35, 41-45, 47, 54
Society of Boiler Makers, 24
Society of Coppersmiths, Braziers and Metal Workers, 38
Society of Steam Engine Makers, 24
South West Counties Vigilance Association, 47
Strachan & Henshaw Ltd., 33, 35

Strikes, 4, 30-32, 34, 36, 38-39, 42, 44-47,
Substitution, See dilution
Supply Departments of the Ministry of Munitions, 14
The Armistice, 43, 48-50, 53, 55
The National Amalgamated Furnishing Trades Association, 20, 27
The Red Flag, 17
Trade Card Scheme, 30, 54
Trade unions, 7, 11-12, 20, 24, 34, 36, 41, 47
Trades Union Congress, 13,
Trafalgar Square stop–the–war demonstration, 22
Treasury Agreement, March 1915, 7, 11
U.K. Smiths and Strikers Union, 24
Unemployment, 34, 51, 55
Unitary corporatism, 11
United Pattern Makers Association, 24
Voluntarism, 11
W & T Avery, 53
War Badges, 10
War Medal Certificates, 10
War Munitions Volunteers' Scheme, 9
War Office, 2, 5, 8
War Savings Certificates, 23
Welton,
Western Daily Press,
White & Poppe Ltd., 25-26
Woodworkers, 6, 20, 27, 38, 39, 45
Women:
Female labour, 6, 11-14, 51
Women metal workers, 20, 40
Women's Peace Crusaders, 43
Women woodworkers, 20, 45
Women workers, 4, 6, 10, 13-17, 19-22, 24-31, 33, 35-36, 38-41, 45, 47-49, 51-55
Workers' Union, 4, 7, 15, 20, 24